**SpringerBriefs in Biochemistry
and Molecular Biology**

More information about this series at http://www.springer.com/series/10196

Jordi H. Borrell · Òscar Domènech
Kevin M.W. Keough

Membrane Protein—Lipid Interactions: Physics and Chemistry in the Bilayer

 Springer

Jordi H. Borrell
Institut de Nanociència i Nanotecnologia
Universitat de Barcelona
Barcelona
Spain

Òscar Domènech
Institut de Nanociència i Nanotecnologia
Universitat de Barcelona
Barcelona
Spain

Kevin M.W. Keough
Alberta Prion Research Institute
Alberta Innovates Bio Solutions
Edmonton
AB
Canada

and

Department of Biochemistry
Memorial University of Newfoundland
St.John's
NL
Canada

ISSN 2211-9353   ISSN 2211-9361   (electronic)
SpringerBriefs in Biochemistry and Molecular Biology
ISBN 978-3-319-30275-1   ISBN 978-3-319-30277-5   (eBook)
DOI 10.1007/978-3-319-30277-5

Library of Congress Control Number: 2016933217

Printed on acid-free paper

This Springer imprint is published by Springer Nature
The registered company is Springer International Publishing AG Switzerland

Foreword

This book is primarily intended for two kinds of audiences: students from biological sciences who desire a quick entry into the area of lipid–protein membrane interactions; and, students from physics or chemistry interested in the physical chemistry underpinning lipid–membrane protein interactions. The book has been conceived as an extension for these undergraduate students and as an introduction to the subject for graduate students who want to enter into membrane research. The objective is to provide a basic background in the physicochemical principles and experimental approaches that are particularly relevant to membrane science. In the first two chapters, we cover elementary topics on lipid and protein biophysical chemistry, self-segregated structures, and experimental methods currently used in membrane research. Although several membrane proteins are mentioned, we chose lactose permease from *Escherichia coli* to introduce most of our experimental examples on lipid–protein structures. Chapters 3 and 4 review the most widespread ideas that have emerged from experimental evidence on lipid–protein interactions. Hydrophobic matching, in which integral membrane proteins induce lipids of the bilayer to adjust hydrocarbon thickness to match the length of the hydrophobic surface of the protein, and the curvature stress lead to the introduction of the so-called surface flexible model for biomembranes. In chapter 4 we provide several examples of correlations of physicochemical properties of lipids with membrane structure, correct folding and function. For methods to assess lipid–protein interactions (DSC, EPR, X-ray crystallography, electron microscopy), the reader is directed to specific references listed at the end of the chapter. The relatively new technique of using AFM-single-molecule force spectroscopy to investigate lipid protein interactions is introduced in Chap. 4.

The purpose of this book is to provide students with an introduction to the physical chemistry of lipid-protein interactions to enable them to extend their studies in this field.

Acknowledgments

We wish to thankfully acknowledge the comments and suggestions of the following people, who have reviewed the manuscript at various stages during its production: Dr. Luis M. Loura, University of Coimbra; and Dr. M. Teresa Montero, Dr. Cristina Minguillón, Dr. Antoni Morros, and Dr. Javier Luque, University of Barcelona: Dr. Manuel Prieto, Centro de Química-Física Molecular, Lisboa for the development of the FRET formalism reviewed in this book: Dr. Mikhail Bogdanov and Dr. William Dowhan from the University of Texas; Dr. Pierre-Emmanuel Milhiet from the Centre de Biologie Structurale de Montpellier; and, Dr. Daniel J. Müller from the ETH of Zürich for figures and AFM images from their work. Thanks are also due to our editor, Thijs van Vlijmen at Springer-Verlag, for the patience and the encouraging comments.

Special thanks go to Dr. Lee D. Leserman, Centre de d'Immunologie de Marseille, Luminy (France), for the privilege of working closely in his lab, long time ago and to Dr. Ronald H. Kaback, University of California Los Angeles (USA), for teaching the lessons learnt from lactose permease.

Barcelona (S)
Edmonton (CDN)
December 2015

Contents

Acronyms

AFM	Atomic Force Microscopy
ATR-FTIR	Attenuated Total Internal Reflectance Infrared Spectroscopy
BAM	Brewster Angle Microscopy
BDH	D-β-Hydroxybutyrate Dehydrogenase
Br	Bacteriorhodopsin
CMC	Critical Micellar Concentration
DSC	Differential Scanning Calorimetry
EM	Electron Microscopy
EPR	Electron Paramagnetic Resonance
FCS	Fluorescent Correlation Spectroscopy
FF-EM	Freeze-Fracture Electron Microscopy
F-MMM	Fluid Mosaic Membrane Model
FRET	Förster Resonance Energy Transfer
GFP	Green Fluorescence Protein
GUVs	Giant Unilamellar Vesicles
IMPs	Integral Membrane Proteins
LacY	Lactose Permease of *E. coli*
LBs	Langmuir–Blodgett Films
LPR	Lipid-to-Protein Ratio
LUVs	Large Unilamellar Vesicles
MIP	Maximum Insertion Pressure
MLVs	Multilamellar Vesicles
MPs	Membrane Proteins
PLSs	Phospholipases
^{31}P-NMR	Nuclear Magnetic Resonance of 31 P
PMPs	Peripheral Membrane Proteins
Rho	Rhodopsin
SFM	Surface Flexible Model
SLBs	Supported Lipid Bilayers
STM	Scanning Tunneling Microscopy

SPMs	Scanning Probe Microscopies
SUVs	Small Unilamellar Vesicle
TEM	Transmission Electron Microscopy
XPS	X-ray Photoelectron Spectroscopy
WLC	Worm-Like Chain Model

Symbols

B	Curvature Elasticity of a Bilayer
c_o	Intrinsic Curvature
C_s	Compressibility of a Monolayer
C_s^{-1}	Compressibility Modulus of a Monolayer
F_u	Unfolding Force
H_I	Hexagonal Phase Type I
H_{II}	Hexagonal Phase Type II
k	Rate Constant
K_A	Surface Compressibility of a Bilayer
K_i	Equilibrium/Association Constant
K_B	Modulus of Compressibility of a Bilayer
L_c	Liquid Condensed Phase
L_{cl}	Contour Length
L_e	Liquid Expanded Phase
L_α	Fluid Phase
L_β	Gel Phase
L_o	Ordered Phase
L_p	Hydrophobic Length
L_L	Hydrophobic Thickness
L_{ch}	Length Chain
M	Bending Moment
r	Fluorescence Anisotropy
r_o	Radius of Curvature
S	Order Parameter
S_o	Equilibrium Surface Area
T_m	Transition Temperature
p	Persistence Length
$p(z)$	Lateral Pressure Density
γ^w	Surface Tension of Water
γ^w	Surface Tension of a Monolayer

κ	Boltzmann Constant
κ_o	Bending Modulus
π	Surface Pressure
η	Viscosity
ν	Molecular Volume
ζ_L	Persistence Length
μ_i	Chemical Potential
χ_i	Mole Fraction

Chapter 1
Molecular Membrane Biochemistry

Abstract In this chapter, a description of the present view of the structure of the cell membrane is presented. This includes a basic introduction to the chemistry and physics of lipids and proteins with special attention to properties that are relevant to understanding lipid-protein interactions within the membrane. It covers protein folding in membranes by describing the interaction forces involved in the process and by focusing on known cases where lipids are involved. By revisiting the fluid mosaic model for membranes the chapter finishes with a visual description of membrane structure at the nanoscale level.

Keywords Lipids · Proteins · Membrane   proteins · Folding · Hydrophobic effect · Membrane structure

1.1 Membrane Architecture

Eukaryotic cells are separated from their environments by their plasma membranes. Several organelles present in the cytoplasm are also surrounded by their own membranes. The primary function of the plasma and intracellular membranes is compartmentalization. That is, they act as physical boundaries for the necessary separation between their respective inner media and their external surroundings. The plasma membrane allows the cell to keep its internal physiological conditions constant in the face of variations occurring in the external medium. Membranes play a crucial role in many physiological and pathological processes: signal transduction, transport of drugs and metabolites, energy generation and the development of tissues, including tumor metastasis, and viral and bacterial infections, among many others. In addition to their plasma membranes, most bacteria and plants present a rigid outer cell wall with a composition and structure that is substantially different from those of the plasma and intracellular membranes. Plasma membranes are constituted by two fundamental building blocks, lipids and proteins. The conventional picture of the biomembrane structure consists of two

© The Author(s) 2016   1
J.H. Borrell et al., *Membrane Protein—Lipid Interactions: Physics and Chemistry in the Bilayer*, SpringerBriefs in Biochemistry and Molecular Biology, DOI 10.1007/978-3-319-30277-5_1

apposed leaflets formed by phospholipids and other lipids (a lipid bilayer) wherein proteins are adsorbed or embedded.

The currently accepted model of membrane structure evolved through different experimental approaches starting from an earlier model which described the membrane as a phospholipid bilayer sandwiched between two layers of globular proteins (Danielli and Davson 1935). This model was itself developed taking into account earlier experiments on the surface activity of membrane extracts (see Sect. 2.1), where it had been deduced that the area occupied by the lipids extracted from erythrocytes was equivalent to twice the total cell surface area (Gorter and Grendel 1925). The current view of the plasma membrane, captured in the fluid-mosaic membrane model (F-MMM) (Singer and Nicolson 1972) is that it is a heterogeneous and anisotropic system that displays remarkable lateral segregation of its components (Fig. 1.1). Such lateral heterogeneity may have different origins. On one hand, biological membranes contain a complex mixture of lipid species that, in the presence of local physico-chemical properties such as pH, temperature, ionic species, or ionic strength may undergo phase separation and lateral segregation into domains of micro- and nanometer dimensions with distinct lipid compositions. On the other hand, some membrane proteins (MPs) may become laterally segregated in lipid domains. This is the case in the spontaneous self-segregation of rhodopsin into crystalline arrays in retinal disc membranes. MP clustering can also be mediated by the specific binding with multivalent peripheral proteins, which is seen in the transmembrane protein band 3 and the peripheral protein spectrin in the erythrocyte membrane or in clathrin-coated pits. Another route to heterogeneous distributions of proteins within a membrane is sustained by intermolecular interactions between specific lipids or groups of lipids and specific proteins that lead

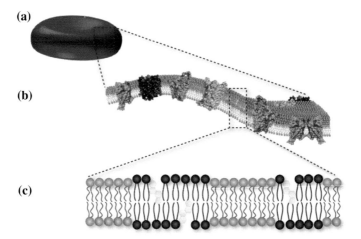

Fig. 1.1 Cartoon illustrating a red blood cell (**a**), an updated version of the fluid mosaic model of a biomembrane displaying proteins adsorbed or embedded in the phospholipid bilayer (**b**), and a membrane detail showing phospholipid regions that are laterally segregated into domains (**c**). (Picas et al. 2012). Reprinted with permission from Elsevier Science

to relatively stable complexes in the plane of the bilayer. A particularly interesting case of lipid-protein segregation in eukaryotic cells is the formation of structures known as "rafts". Rafts are conceived as dynamic platforms of special lipid and protein composition into and out of which proteins would diffuse, allowing for various protein–protein contacts that might modulate associated protein functions such as signal transduction. Lipids in direct contact with proteins that extend through membranes are sometimes called annular or boundary lipids (see Chap. 3). Domains such as rafts are considered to contain more lipids than only boundary lipids. Despite much experimental evidence supporting the existence of raft domains, and boundary or annular phospholipids, controversies about their existence remain because such structures have not been definitively demonstrated by means of direct visualization techniques.

Although the basic structure of the membrane is considered to be universal, the diversity in cellular morphology indicates that there is a rigorously defined composition and organization that is associated with the particular function of each kind of cell or organelle.

1.2 Chemistry and Physics of Membrane Lipids

The lipids found in cell membranes belong to four major groups: glycerophospholipids, sphingolipids, glycolipids and sterols. Although they share 3-dimensional structural similarities, glycero- and sphingolipids are constructed with different backbones, glycerol and sphingosine. As seen in Fig. 1.2 one or two long chain hydrocarbons are esterified in each lipid type.

Glycolipids have sugars such as glucose, mannose or inositol attached to glycerol or sphingosine in addition to the long hydrocarbon chains.

In glycerophospholipids, the glycerol backbone is esterified to acyl chains in position 1 (R1) and 2 (R2), and to a phosphate group in position 3 to form phosphatidic acid. Notice that with sphingosine, one of the acyl chains is linked to an

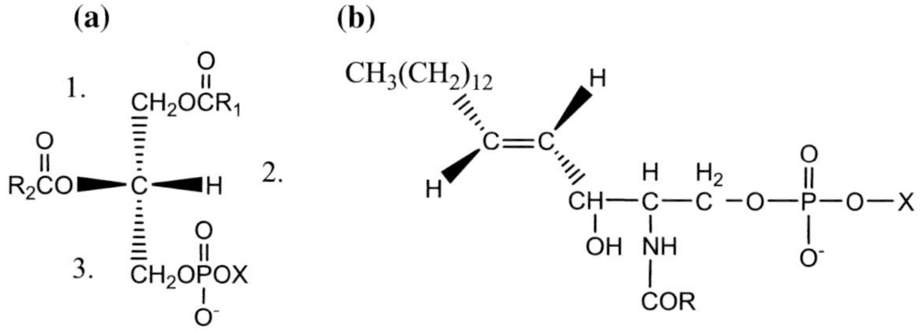

Fig. 1.2 Greenall structure for phospholipids: glycerophospholipids (**a**); sphingolipids (**b**). According to the Fisher projection, the secondary hydroxyl (C-2) group of glycerol is drawn on the left, the carbon above is called C-1 and the one below C-3

amino group. The numbering of the carbons denotes the chirality of the glycerol which gives rise to stereoisomers. In natural membranes, all glycerophospholipids show the L-absolute configuration. The acyl chains R1 and R2 can be saturated or unsaturated, with more common lengths from C16 to C20.

The most common acyl chains found in animal cell membranes are saturated palmitic (16:0) and stearic (18:0) acids, and unsaturated oleic (18:1), linoleic (18:2) and arachidonic (20:4) acids (Table 1.1). When a phospholipid contains two identical acyl chains it is referred to as a homoacid lipid; if the chains are different the name heteroacid lipid is used. Most naturally-occurring lipids are heteroacid lipids such as in the case of mammalian lipids which contain a saturated and an unsaturated lipid chain. In phospholipids the phosphate groups can be esterified to the hydroxyl groups of bases such as choline, serine, ethanolamine or inositol, leading to phosphatidylcholine (PC), phosphatidylethanolamine (PE), phosphatidylserine (PS), phosphatidylglycerol (PG) and phosphatidylinositol (PI), respectively (Table 1.2).

Table 1.1 Names and abbreviations of the most common glycerophospholipids

-X in Fig. 1.2a	Generic name	Name	Abrr
–H	Phosphatidic acid	1,2-diacyl-sn-glycero-3-phosphoric acid	PA
–CH$_2$–CH$_2$-N$^+$(CH$_3$)$_3$	Phosphatidylcholine	1,2-diacyl-sn-glycero-3-phosphocholine	PC
–CH$_2$–CH$_2$-N$^+$H$_3$	Phosphatidylethanolamine	1,2-diacyl-sn-glycero-3-phosphoethanolamine	PE
–CH$_2$–CH(COO$^-$)-N$^+$H$_3$	Phosphatidylserine	1,2-diacyl-sn-glycero-3-phosphoserine	PS
–CH$_2$–CHOH-CH$_2$OH	Phosphatidylglycerol	1,2-diacyl-sn-glycero-3-phosphoglycerol	PG
	Phosphatidylinositol	1,2-diacyl-sn-glycero-3-phospho-1′myo-inositol	PI

Table 1.2 Names and notation of common fatty acids found in membranes. The first number of the symbol refers to the length of the acyl chain and the second to the number of double bonds. The positions of the double bonds are noted in parentheses

Common name	Symbol	Common Name	Symbol
Myristic	14:0	γ-Linoleic	18:3 (6, 9, 12)
Palmitic	16:0	α-Linoleic	18:3 (9, 12, 15)
Palmitoleic	16:1	Arachidic	20:0
Stearic	18:0	Arachidonic	20:4 (5, 8, 11, 14)
Oleic	18:1 (9)	Behenic	22:0
Linoleic	18:2 (9, 12)	Lignoceric	24:0

Obviously there will be a huge number of structures depending upon the combination of hydrophobic chains, sugars and bases incorporated into the molecules. The nomenclature is tedious but important, particularly in research papers, where the precise nature of the lipids used in experimental approaches using model membranes must be carefully identified. Thus, some examples of the lipids that dominate in the literature are shown in Table 1.3.

DPPC (1,2-dipalmitoyl-sn-glycero-3-phosphocoline) is the most abundant phospholipid in mammalian lung surfactant (see Chap. 2). It is also the molecule most widely used by researchers to prepare model membranes (monolayers and liposomes). The prefix *sn* is distinguished to specify the stereochemistry of the compound and the prefix *rac* (i.e. shown in POPG) is used to indicate that the compound is a racemic mixture. POPE (1-palmitoyl-2-oleoyl-sn-glycero-3-phosphoethanolamine) and POPG (1-palmitoyl-2-oleoyl-sn-glycero-3-[phospho-rac-(1-glycerol)]), are the predominant phospholipids found in gram-negative bacteria.

Another large group of hydrophobic compounds found in membranes are the steroids. Sterols containing a hydroxyl group which is capable of being esterified (Fig. 1.3) are most common. The most frequently occurring sterol is cholesterol (CHOL) which is present in mammalian cells, but it is absent from the membranes of prokaryotes. In higher plantes, β-sitosterol is the major sterol. Other sterols as stigmasterol and ergosterol are found in higher plants and yeast. The major effect of CHOL in membranes is to modify the fluidity of the membrane in a concentration-dependent manner, conferring optimal physical properties for the different physiological functions. Often, membranes depleted of cholesterol loose structural

Table 1.3 Some examples of phospholipids commonly encountered in model membranes research

Phospholipid	Abbr.	Acyl chain composition	Structure
1,2-dipalmitoyl-sn-glycero-3-phosphocoline	DPPC	Homoacid (R1=R2=C:16)	
1-palmitoyl-2-oleoyl-sn-glycero-3-phosphoethanolamine	POPE	Heteroacid (R1=C:16; R2=C:18:1Δ9)	
1-palmitoyl-2-oleoyl-sn-glycero-3-[phospho-rac-(1-glycerol)]	POPG	Heteroacid (R1=C:16; R2=C:18:1Δ9)	
1,2-dioleoyl-sn-glycero-3-phosphoethanolamine	DOPE	Homoacid (R1=R2=C:18:1Δ9)	

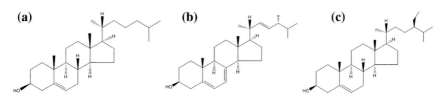

(a) **(b)** **(c)**

Fig. 1.3 Structures of **a** Cholesterol, **b** ergosterol and **c** β-sitosterol

Wintegrity and become highly permeable. Although CHOL is a four-ring hydrocarbon planar molecule, the hydroxyl group in C3 confers on it an amphipatic
nature that enables it to align with the hydroxyl region at a bilayer interface with
the ring structure embedded within the hydrophobic region bilayer. The hydroxyl
may interact with water and the polar moiety of phospholipids.

All lipids in the membrane have an amphipatic nature, consisting of a hydrophilic
"head group" of varying polarity, and a hydrophobic "tail" consisting of the hydrocarbon portion which "avoids" water through being sequestered in the membrane or
bilayer interior. At neural pH, the phosphate and amino groups of the polar head are
both charged, and so PC and PE behave as zwitterions. Other lipids, such as PG,
PS, and PI, usually carry a net negative charge at neutral pH and less frequently a
positive charge, as is the case for 1,2-dipalmitoyl-sn-glycero-3-ethylphosphocholine
(EDPPC or Et-DPPC). As we will discuss in Sect. 2.5 the balance between "head"
and "tails" will determine the formation of self-organized structures.

1.3 Membrane Proteins

1.3.1 Protein Structure

Proteins are macromolecules consisting of amino acids as their monomeric building blocks. The 20 amino acids most commonly occurring in nature share a general structure with an α carbon atom (C_α) adjacent to a carboxyl (-COOH) group,
with the other positions ocupied by an amino (-NH$_2$) group, a hydrogen (H) atom
and a variable side chain (-R) group.

The C_α adopts a tetrahedrical disposition, and since the four substituents are all
different, amino acids are chiral. Hence (with the exception of glycine) two stereoisomers exist of each amino acid, each one a non-superimposable mirror image
of the other (Fig. 1.4). Proteins are synthesized on the ribosomes using information encoded in messenger RNA, a process called translation. Although there are
20 genetically codified amino acids (shown in Figs. 1.5 and 1.6), very often structural variations of them can be found due to postransductional modifications. All
naturally occurring proteins are composed of amino acids with L-absolute configuration (D/L, Fisher's nomenclature).

Amino acids are mainly classified according to the physico-chemical properties of the chemical moiety that form the side chains. At a high level we can
classify the amino acids as polar or apolar, depending on their solubility in water.

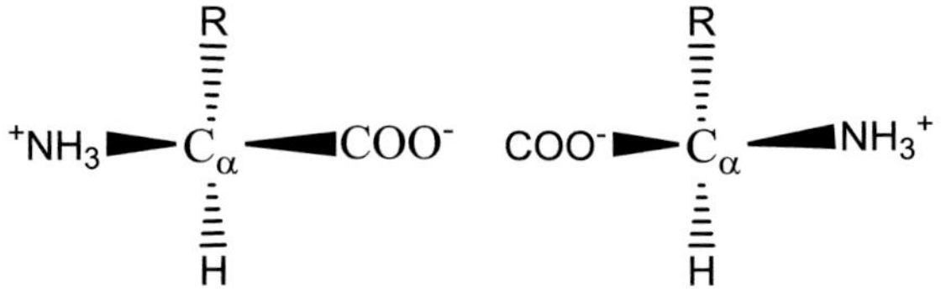

Fig. 1.4 Fisher projection of amino acids

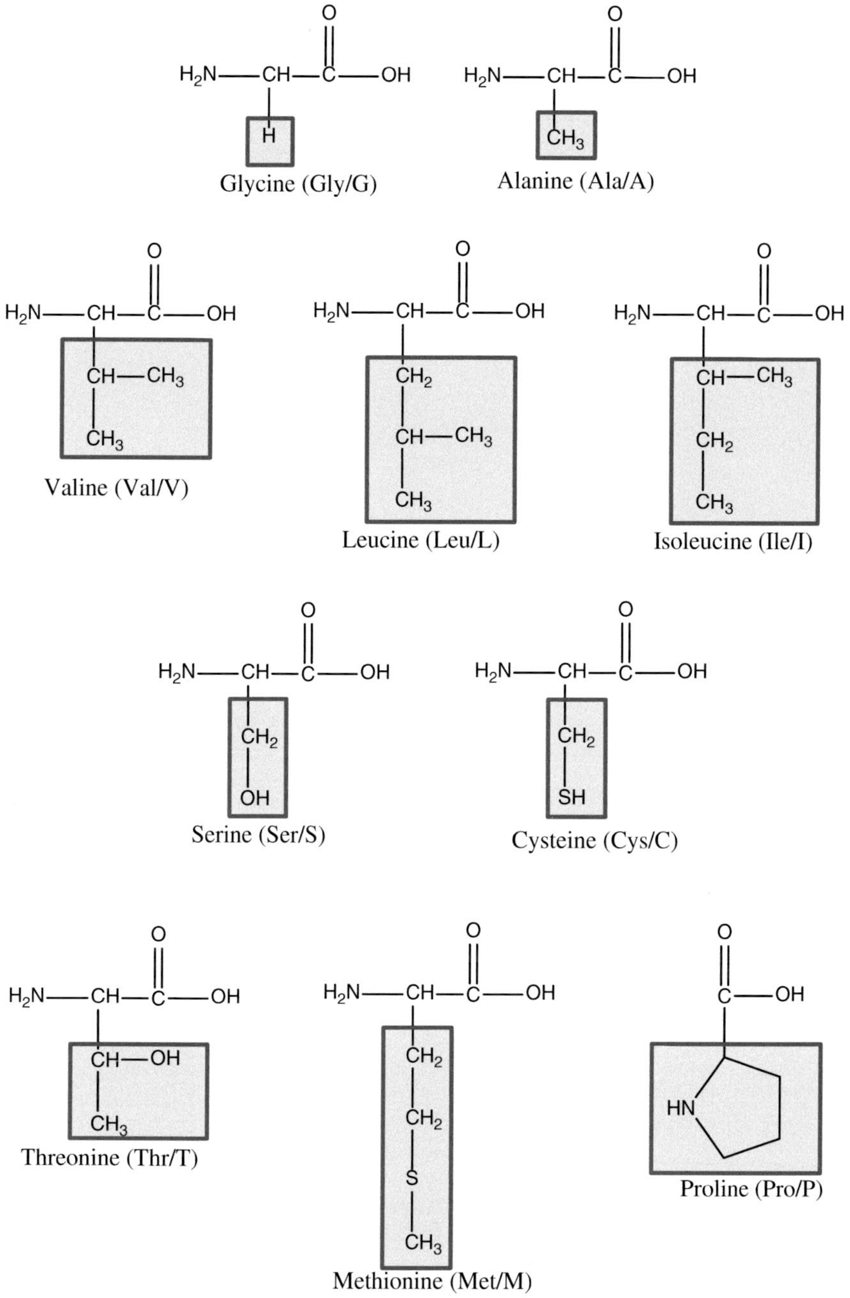

Fig. 1.5   Aliphatic amino acids

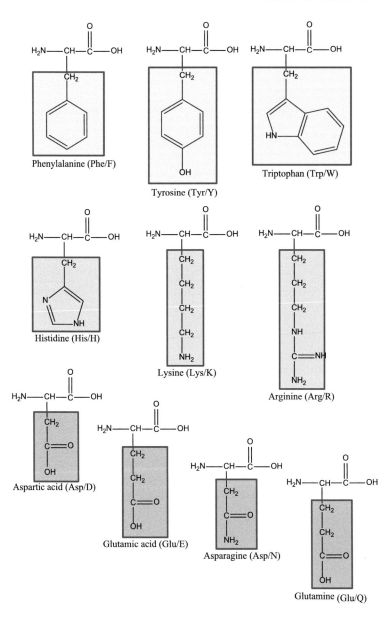

Fig. 1.6 Aromatic, positively and negatively charged amino acids

At a more refined level, amino acids are classified into five categories: (i) nonpolar aliphatic (Fig. 1.5); (ii) nonpolar aromatic; (iii) polar uncharged; (iv) positively charged; and (v) negatively charged (Fig. 1.6).

The smallest amino acid is glycine (G) for which –R═–H. Because there are two identical H substituents in C_α, this amino acid does not have stereoisomers. Due to its small size, G confers flexibility on proteins, providing room for movement of neighboring amino acids. The nonpolar amino acids with hydrophobic side chains are alanine (A), valine (V), leucine (L), isoeucine (I) and the imino acid proline (P). Although methionine (M) contains a sulphur in its side chain its properties are also hydrophobic and nonpolar. Proline is a rigid amino acid that cannot participate in hydrogen bonding and this property plus its three dimensional shape cause proline to be a disruptor of the secondary structure of the protein. The other non polar amino acids confer certain rigidity to the protein, hence they are considered "structural", which means that they form part of the non specific backbone of a protein. M, through its S atom, participates in some enzymatic reactions.

The nonpolar aromatic group is formed by phenylalanine (F), tyrosine (Y) and tryptophan (W). These amino acids are natural fluorophores (Table 1.4). In fact, protein fluorescence arises mostly from the W residues, which predominate over the fluorescence of F or Y. As can be seen by a simple inspection of Table 1.4, W presents much stronger fluorescence and a higher quantum yield than the other two aromatic amino acids. Actually, a protein becomes fluorescent in the presence of only one residue of W. The fluorescence of W can be exploited, for instance, to investigate the polarity of the environment, in quenching experiments (see Sect. 4.2) or in fluorescence methods based on the Förster resonance-energy transfer mechanism (FRET) (see Sect. 4.9).

Among the polar and neutral amino acids, asparagine (N) and glutamine (Q) have polar amide groups, and threonine (T) and serine (S) have polar hydroxyl groups. These amino acids are able to establish hydrogen bonds with other polar molecules, therefore being key in the formation and maintainance of the tertiary structure of the protein. Cysteine (C), in turn, contains a sulfhydryl (-SH) group, which can be oxidized to form a disulfide (-S-S-) bond. The formation of disulfide bridges stabilizes proteins, but requires an oxidative environment, found in extracellular media. Since the –SH group of C is very reactive; it has been exploited for protein covalent labeling for spectroscopic purposes (i.e. fluorescence or spin resonance spectroscopy). S and T, due to the nucleophilic activity of –OH and –SH, are often active sites in the reaction center of many enzymes.

Aspartic (D) and glutamic (E) acids, the polar amino acids, are normally present as aspartate and glutamate, bearing a net negative charge at neutral pH. Together with the positively charged arginine (R), and lysine (K), these amino acids are responsible for the overall charge of a protein. Histidine (H) contains an imidazole ring as a side chain. Since its pKa is 6.8, a small shift to a lower

Table 1.4 Typical fluorescence characteristics of fluorescent aminoacids. Maximum wavelengt of excitation (λ_{max}^{exc}), molar absortivity (a_M), maximum wavelength of fluorescent emission (λ_{max}^{em}), quantum yield (ϕ_f) and fluorescence lifetime (τ_f)

Amino acid	λ_{max}^{exc}(nm)	a_M (M^{-1} cm^{-1})	λ_{max}^{em}(nm)	ϕ_f	τ_f (ns)
Tryptophan	280	5600	348	0.20	2.6
Tyrosine	274	1400	303	0.14	3.6
Phenylalanine	257	200	282	0.04	6.4

pH provokes the protonation of imidazole, changing it into a polar and positively charged amino acid. H is involved as a catalyst in many enzymatic reactions.

The primary structure of a protein is originated by the formation of peptide bonds between amino acids in a linear arrangement, or the sequence that constitutes a polypeptide chain. This particular bond is formed by a condensation reaction between the $-NH_2$ group of one amino acid and the $-COOH$ group of another (Fig. 1.7). The resulting bond is an amide, the peptide bond, which has limited rotation around the $O=C-N$ bond. This is a consequence of electron delocalization over the $O-C-N$ system that confers double-bond characteristics with restricted rotation upon the C-N link. A polypeptide or a protein is a succession of peptide units that progress from a C_α to the following C_α in the amino acid sequence. Between every two C_α, except for the initial and the final ones, the peptide unit constitutes a rigid plane (Fig. 1.8). Therefore, every C_α is shared by two peptide planes that can rotate around the adjacent C_α-C' and N-C_α bonds. The corresponding angles of rotation that define the relative orientation between two consecutive planes are normally labeled as phi (ϕ) and psi (ψ), respectively.

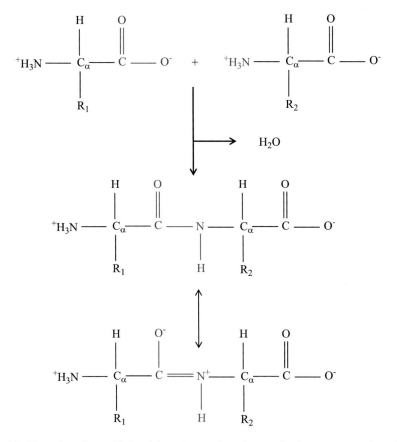

Fig. 1.7 Formation of a peptide bond through a condensation reaction between two amino acids

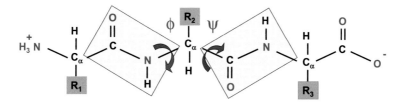

Fig. 1.8 Schematic depiction of a peptide backbone, illustrating the dihedral angles ϕ and ψ. In a polypeptide the backbone chain can freely rotate around N-C_α (ϕ) and C_α-C (ψ) angles

The conformation angles, however, cannot adopt all values. Analysis of the molar volumes of the amino acids provides an intuitive basis for the understanding that most combinations of ϕ and ψ will be not allowed because steric collisions between the side chains and main chain might occur. The polypeptide will be fully extended when ϕ = ψ = + 180°. When ϕ = ψ = 0° the two consecutive planes will adopt a sterically impeded conformation because there will be an atomic contact between the oxygen from the first plane and the hydrogen from the amide group in the second plane. When ϕ = 180° and ψ = 0°, restrictions in rotation between the planes will occur because of steric hindrance between the hydrogens and the peptidyl nitrogens. When ϕ = 180° and ψ = +120°, the C_α-H bond is trans to the C=O bond. When ϕ = 120° and ψ = +180°, the C_α-R bond is trans to the N-H bond. When pairs of ϕ and ψ are plotted against each other, the so-called Ramachandran plot (Ramachandran et al. 1963) is constructed by using all possible values of ϕ and ψ (Fig. 1.9). The diagram is useful in order to visualize

Fig. 1.9 Ramachandran plot. The white areas correspond to conformations where atoms in the polypeptide come closer than the sum of their van der Waals radii. Blue and pink areas correspond to sterically allowed conformations α, β and C

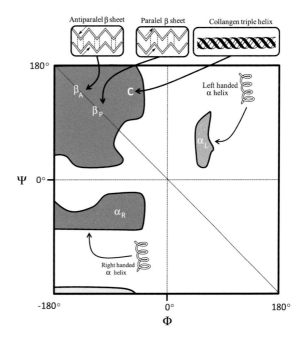

energetically unfavorable conformations, particularly when constructing 3D structural models of proteins.

When a macromolecule is formed by monomers that rotate freely around a defined bond, and do not interact between themselves in an orderly fashion, the monomer chain is called a random coil. This type of structure is apparent in proteins. In truth the random coils found in proteins are not truly random because the primary sequence of the various amino acids drives the structure with similar interactions between some amino acids each time the protein is formed during protein synthesis. Several side-chain interactions arise that lead to higher levels of organization than the primary structure, as defined by the linear arrangement of the amino acids. Thus, in addition to random coils, two other different stable structures exist: α-helixes and β-sheets. These, together with random coils, are the secondary structures that might be present in a single protein.

In an α-helix, a hydrogen bond is formed between the C=O group of one peptide group and a NH group in the peptide group three residues further along the chain towards the C-terminus. The structure is doubly stabilized, because a second hydrogen bond is formed between the amide hydrogen of the amino acid and the C=O group of the residue three positions backwards towards the N-terminus. α-helices have 3.6 amino acids per turn, with an axial rise per residue of 0.15 nm along the axis and values for all the amino acids of $\phi \sim -60°$ and $\psi \sim -50°$. The α-helix is a very regular and stable arrangement, with all the hydrogen bonds pointing along the helical axis. Because each peptide unit displays a dipole moment that is roughly aligned parallel to the helix axis, there is a net dipole moment for the overall α-helix, with positive and negative poles at the –NH and –COO terminal groups of the helix, respectively.

Amino acids like A, E, L or M best fit the requirements for the α-helix structure at the other end of the spectrum are P, G, Y or S which are much less suitable for and α-helix structure. P introduces a distortion of ~20° relative to the axis, due to steric hindrance arising from its cyclic side chain which prevents the main-chain N atom from participating in hydrogen-bonding. An α-helix cannot be sustained in the presence of a proline, so this amino acid is often referred to a as a "helix-breaker". A different force causing distortion in α-helices originates from the tendency of C=O groups to point to the bulk solvent in order to optimize hydrogen bonds with water.

The other main secondary structure found in natural occurring proteins is the β-sheet. This structural motif is characterized by angles of $\phi \sim -140°$ and $\psi \sim +130°$. This means that polypeptides in a β-sheet display an extended conformation. The β-sheet structure occurs when hydrogen bonds between the main chain C=O and N-H groups are formed between two different β strands that are aligned side by side, with the side chains of the residues pointing in opposite directions. The polypeptide chains are arranged in sheets formed by adjacent chains that give n the sheets a series of alternating ridges and valleys, and are described as a pleated sheet. β-sheets can be classified as parallel or anti-parallel, depending on whether the strands run in one direction or whether they run alternately in opposite directions. Such parallel and anti-parallel sheets can be present is the same protein.

Integral membrane proteins usually consist of transmembrane α-helical segments that are connected by cytoplasmic and extracellular loops. The extramembraneous loops usually have unique secondary and tertiary structures that are involved in the protein function and contribute to the stability of the whole protein.

Individual proteins contain many combinations of different amounts of α- and β-motifs plus contributions from random coils. The overall three-dimensional shape conferred on the protein by the combination of those elements is called its tertiary structure. A protein can present an extended rodlike shape, named fibrous or adopt a spherical or ellipsoidal shape, termed globular. Many proteins consist of more than one polypeptide chain that are held in association through non-covalent bonds. The three-dimensional arrangements between the associating units are termed the quaternary structure of the protein. The protein folding that leads to the higher levels of structure is the result of the intermolecular interaction resulting in most cases from non-covalent interactions between amino acid side chains (e.g., hydrogen bonds, ionic pairs or hydrophobic interactions).

1.3.2 Membrane Protein Structure

Due to the amphipathic nature of the phospholipids, three regions can be distinguished in a membrane: two polar regions at the surface of each side of a bilayer, separated by a hydrophobic region constituted by the acyl chains (Fig. 1.10). Membrane proteins (MPs) are usually classified in two main categories: integral (IMPs) and peripheral (PMPs), depending on whether the protein crosses the membrane from one side to another or not, respectively. Although the characterization implies the structural location of the proteins, their assignment is often based on their behavior in processes intended to extract them from a membrane. While PMPs are easily extracted from the membrane by using high ionic strength buffers, IMPs (much more hydrophobic than PMPs) require the use of organic solvents or detergents. Since variation of the ionic strength changes the degree extent of PMP extraction, it can be deduced that the interaction between these proteins and the membrane are primarily of an electrostatic nature. IMPs must be amphiphilic in nature, with hydrophobic regions that span the bilayer from once to many times, and hydrophilic regions in contact with either the cytoplasm or the extracellular space. The majority of the hydrophobic residues are hidden within the protein or in contact with the acyl chains of the phospholipids, and the charged residues can be exposed to the cytoplasmic or external aqueous media. Since natural membranes often present a negative charge conferred by acidic phospholipids, specific interactions with positively charged residues of proteins are possible. The requirement of organic solvents or detergents for extracting IMPs provides evidence for the apolar nature of these proteins. Glycophospholipid-linked proteins are often also considered an independent class of MPs.

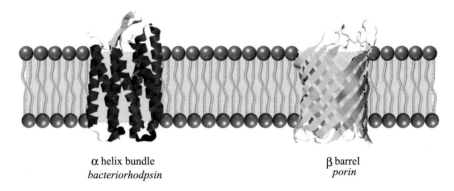

α helix bundle
bacteriorhodpsin

β barrel
porin

Fig. 1.10 Structures of membrane proteins α-helical bundle and β-barrel proteins

IMPs are transmembrane proteins constituted predominantly by membrane-spanning α-helices that assume a specific orientation with respect to the membrane normal. The helices are connected by hydrophilic loops of the protein that are located in the aqueous space on either side of the bilayer(Fig. 1.10). IMPs containing β–strands that span the membrane and form a β-barrel tertiary structure are less common in most eukaryotic membranes but they are found in a significant number of proteins in bacterial outer membranes. It can be assumed that α-helices will require more residues (~20 residues) than β-barrels (~10 residues) per transmembrane unit . However, this is not an absolute physical requirement, among other reasons because of the presence of hydrophilic residues in the transmembrane strands or distortions imposed by lipid arrangements around the IMP.

In soluble proteins, whereas the protein surface is populated by hydrophilic residues, the hydrophobic residues are kept apart from water in the interior core of the protein. This is a principle that is applicable to PMPs. In IMPs the surface of the protein facing the bilayer core will be populated by hydrophobic residues, whilst charged residues will predominate in the connecting loops. In some instances hydrophilic cavities occur in the interiors of proteins with multiple membrane-spanning regions which contain inward facing hydrophilic residues. Some residues such as K and R are more frequent in hydrophilic loops at the cytoplasmic face, providing a net positive charge in this area of the protein. This is known as the "positive-inside rule" for membrane proteins. It is interesting to note that aromatic residues, particularly W, have some preference for the interfacial region (White et al. 1998).

As discussed above, β-sheets require lateral hydrogen bonding between the different strands. Therefore, the polypeptides must cross the bilayer several times in order to form a stable β-sheet conformation. A large amount of IMPs with β-sheets have been found in the outer membranes of bacteria, mitochondria, and chloroplasts. Porins are examples of transmembrane proteins with 16-18 β-strands whose structure looks like a cylinder or β-barrel. Porins are assembled in trimers, each monomer with a pore like-channel with a diameter of ~1.5 nm, through which a wide variety of polar molecules are transported in a non-selective way. The side chains of the residues facing the inside of the pore are hydrophilic. Conversely, the residues facing the bilayer are predominantly hydrophobic. It should be mentioned that, with this assembly, the hydrogen bonding requirements of the polypeptide chains are satisfied within the membrane.

As mentioned above, the predominant motif found in IMPs is the α–helix. The first sequenced protein of this kind was glycophorin, which is present in erythrocytes. Glycophorin is made up of three parts: the cytoplasmic, intramembrane α-helical and extramembrane domains. Glycophorin, like the VSV-G protein or the insulin receptor, has a single transmembrane α-helix. However, most IMPs contain multiple membrane spanning α-helices. Widely investigated IMPs include bacteriorhodopsin (Br), ubiquous in *Hallobacterium* species, that folds into a seven transmembrane helix topology; the bacterial photosynthetic reaction center, with eleven transmembrane α-helices; and the lactose permease (LacY) of *Escherichia coli* (Fig. 1.11), with 417 amino acid residues and twelve transmembrane α-helices, is a paradigm for secondary active transporters that use the membrane electrochemical potential ($\Delta\tilde{\mu}_i$) as the energy source to transport solutes. Much is known about the structure and function of LacY, with 70 % of its amino acid residues located in hydrophobic domains. LacY consists of twelve transmembrane α-helices that are connected by eleven relatively hydrophilic, periplasmic

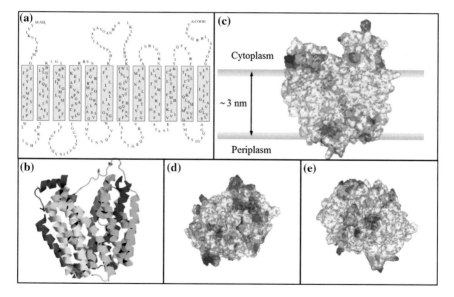

Fig. 1.11 **a** Secondary structure of lactose permease from *E. coli* showing the charged residues (*bold*). The one-letter amino acid code is used. The putative transmembrane helices are shown in *boxes* that are connected by hydrophilic loops; **b** cartoon 3D view of lactose permease; **c** surface model and electrostatic potential of lactose calculated with the program Protein Explorer (Martz 2002). The polar surfaces are colored *blue* (positively charged) and *red* (negatively charged); **d** view along the membrane normal from the cytoplasmic side; **e** and periplasmic side. Reprinted from Merino-Montero et al. (2006) with permission from Elsevier Science

and cytoplasmic loops, with both N- and C-terminal groups on the cytoplasmic surface (Fig. 1.11a). Viewed along the membrane axis (Fig. 1.11d), LacY is heart shaped, and presents an internal cavity that opens to the cytoplasm (Merino-Montero et al. 2006).

Following a general tendency for membrane proteins, the 5 **W** residues of LacY are located predominantly near the interface formed by the phospholipid headgroups. Noteworthy, none of **W** residues are essential for activity, but they are important for insertion and stability in the membrane. Notably, one **W** residue (W151) is situated in the middle of the hydrophilic cavity and plays a key role during substrate recognition and subsequent conformational transitions of the protein.

Many ionizable residues are present in the interconnecting loops, facing the polar cytoplasmic and periplasmic regions, but some are present within the nonpolar regions of the bilayer. This situation, the predominance of charged residues in the connecting loops, could be relevant in regard to a possible association of phospholipids with the protein. This aspect will be discussed in Chap. 4.

1.3.3 Membrane Protein Insertion in Natural Membranes

One of the unresolved issues in membrane biology is the mechanism of insertion of IMPs into the phospholipid bilayer (Cymer et al. 2014).The insertion of the IMPs is a key biological event because the protein must fold not only in its tertiary (or even quaternary) structure, but also into the adequate topology required for its physiological activity. For instance, glycophorin and the transferrin receptor both have a single α-helix transmembrane domain, but whilst the glycophorin –C terminus is exposed to the cytosol (Type I transmembrane protein), it is the N–terminus of transferrin that is directed towards the intracellular medium (Type II transmembrane protein). How transcription results in the adequate orientation of each IMPs, is an intriguing and incompletely understood process that needs to deal with the hydrophobicity of the residues of each protein, the relative position of the different residues along the transmembrane segment and the amphipatic nature of the latter.

Similarly to soluble proteins, MPs are synthesized as polymer chains that are subsequently folded into two dimensional and tertiary structures. It is likely that IMP asymmetry could be initiated at the time of their biogenesis. In IMPs, all the pathways involved in the hierarchical organization take place within the two dimensional medium provided by the membrane. The insertion of the protein in the membrane occurs when the polypeptide chains are secreted by ribosomes into a protein conducting channel called a translocon. Translocons are integral membrane proteins (often associated with other proteins), that act as molecular gatekeepers allowing newly synthesized polypeptide chains to pass across the membrane or integrate into the lipid environment. Assisted by the transposon and

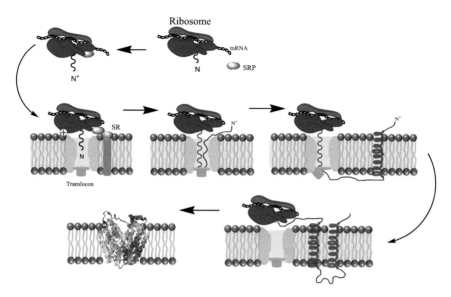

Fig. 1.12 Overview of membrane protein synthesis, membrane insertion and assembly. Adapted from Dowhan and Bogdanov (2009)

other proteins called chaperones, all transmembrane segments are assembled in the membrane to reach the native folded state of the protein (Fig. 1.12). The mechanism by which transposons assist membrane proteins to insert in the membrane is as yet not fully resolved. What it is clear is that, in this process, the influence of the phospholipids is key, for example, in the case of LacY where the phospholipid PE is essential for correct topology, apparently acting as a chaperone for the correct folding of the protein (Bogdanov et al. 1996).

This requirement for PE was discovered by comparing the activity of cells lacking PE with wild-type *E. coli* cells (Dowhan and Bogdanov 2009) and using a conformation-specific monoclonal antibody (4B1) that specifically recognizes the P7 domain, the loop between helix VII and Helix VIII of LacY (Sun et al. 1997). In the absence of PE, LacY shows a different non-native topology. When PE is lacking helices I-VI are in an "inverted" orientation with respect to helices VIII-XII in comparison to the orientations in the native topology (see Fig. 4.2). In addition helix VII is exposed completely to the periplasm, resulting in the disruption of the salt bridge between helix X and helix XI. Importantly, once LacY is properly folded in the membrane, PE is no longer required for maintainnance of the protein conformation. In molecular terms the requirement for PE has been demonstrated to be specific for several species. These experiments have been corroborated by the reconstitution of LacY in PE and PG-CL liposomes (Wang et al. 2002). Only PC and PE protoliposomes presented the correct topological organization as judged by the orientation of the P7 domain. All these experiments demonstrate evidence for subtle interplay existing between IMPs and host lipids in membranes.

It is known that apart from disulfide and peptide bonds, protein folding is a complex mechanism that involves non-covalent forces as well, such as ionic interactions, dipole interactions, hydrogen bonds, van der Waals forces and, non-specific hydrophobic interactions. The basic contribution of each kind of interactions is briefly schematized in Fig. 1.13.

Hydrogen bonds mostly involve C=O and N-H groups of the protein backbone and they occur when two electronegative atoms compete for the same hydrogen. Covalent bonds established by oxidation of two cysteine amino acids are also important in the folding process; the most well-known example is the folding of ribonuclease usually described in basic biochemistry textbooks. Ionic interactions also play a role in folding of the peptide chain. They occur mostly between NH_4^+ and COO^- moieties on amino acid side chains. Dipole-dipole interactions and van der Waals interactions also play a role; the first occurring between groups of different eletronegativities and the second due to polarization effects that are weak and of close range with a distance dependency of r^{-6}.

Protein folding can be formulated as

$$P(n) \rightleftarrows P(u) \rightarrow P'(d)$$

where P(n), P(u) and P'(d) represent the natural folded, unfolded and denatured protein, respectively. Whereas the unfolding process is reversible, the denaturation step is irreversible. By assuming a two-state approximation for the unfolding process, the apparent equilibrium constant can be written as

$$K_U = \frac{[P(d)]}{[P(n)]} \tag{1.1}$$

Since the folded state is the more stable, the variation of the Gibbs energy for the folding process ($\Delta_f G$) will be negative, as expected for any spontaneous process. The theoretical folding profile of a protein can be understood by plotting the Gibbs energy as a function of the folding coordinate (Fig. 1.14). The folding coordinate represents a geometrical concept that denotes all possible states reached by the protein when the transition between the two states P(n) and P(u) occur.

Then the stability of the protein can be written as

$$\Delta \Delta_f G = \Delta_N G - \Delta_U G \tag{1.2}$$

where $\Delta_N G$ and $\Delta_U G$ are the Gibbs energies of the native and unfolded state, respectively. Typical values for $\Delta \Delta_f G$ range from -15 to -50 kJ mol^{-1}.

1.3.4 Hydrophobic Effect and Protein Folding

It is generally agreed that one of the main factors which determine the tertiary structure of proteins is a process driven by entropic force which is often called hydrophobic effect. Because of the high hydrophobicity of IMPs, this effect may play a more important role in the insertion and folding of membrane proteins into

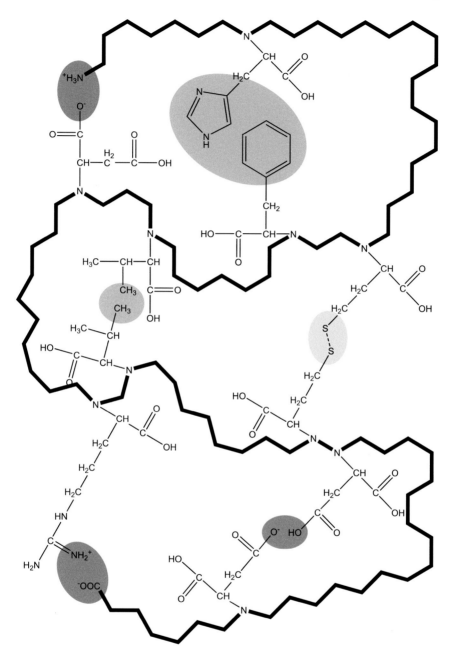

Fig. 1.13 Theoretical force interactions in a polypeptide chain. Ionic interactions (pink); hydrophobic interactions (gray) and sulfide bridges (yellow)

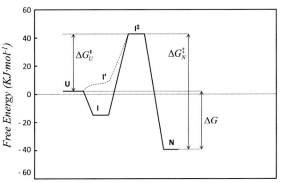

Fig. 1.14 Gibbs energy diagram for the protein folding process, assuming a two-state model. $\Delta G^{k}_{folding}$ and $\Delta_U G^{\perp}$ denote the activation energy between the unfolded and transition states and between the respective native and transition states

the membrane than it does in the maintenance of tertiary structure. This phenomenon reflects a tendency of non-polar molecules to avoid contact with water, and by so the total entropy of the system is reduced. A most familiar observation of a process that is governed by the hydrophobic effect is the limited solubility of non-polar compounds in water, and their coalescence into separate phases such as oil drops within an aqueous environment. This spontaneous interaction is a source of energetic stability, because the unfolding or denaturation of a protein would leave the non-polar residues exposed to the unfavorable aqueous environment. The rationale behind the hydrophobic effect has its origin in the change of the entropy associated with the water molecules. In the unfolded state of a protein, molecules of water are forced to form cage-like structures around the non-polar side chains. This cluster effect imposes a loss of degrees of freedom or entropy for the molecules of water. We can write the Gibbs energy for unfolding as

$$\Delta_u G = \Delta_u H - T\Delta_u S \tag{1.3}$$

hence the decrease in the entropy of water ($\Delta_u S$) will make $\Delta_u G > 0$. Thus, the native folded state of the protein will be always favored (see Fig. 1.14).

The contribution of the hydrophobic effect to the protein stability can be estimated by measuring the Gibbs energy of transfer of each amino acid from organic solvents to water. Quantitative values based on amino acids partitioning between the aqueous phase and an organic phase (i.e. ethanol, octanol, cyclohexane) can be obtained by using the equation

$$\Delta_{transf} G_{aa} = \mu_{aa}^{org} - \mu_{aa}^{aq} = -2.303RT \log \frac{\chi_{aa}^{org}}{\chi_{aa}^{aq}} \tag{1.4}$$

where μ_{aa}^{org} and μ_{aa}^{aq} are the chemical potentials of the amino acid in the aqueous and the organic phase, respectively, and χ_{aa}^{org} and χ_{aa}^{aq} are the mole fractions in each phase. The selection of an appropriate organic solvent should be based on its resemblance to the interior of the protein or the hydrophobic portions of the membrane and, it should not interact with the amino acid side-chains. One of the earliest scales of hydrophobicity was derived by measuring the energy transfer from water to ethanol (Nozaki and Tanford 1971). In order to estimate the contribution of each side chain, the experimental values from each amino acid were subtracted from the transfer energy of lysine. However, the use of ethanol as organic solvent should be taken with caution, because the non-interacting condition is not well fulfilled. The quest for an absolute scale of hydrophobicity continues, and many experimental and theoretical works designed to establish the net contribution of each amino acid side-chain to protein folding, can be found in the literature. Some of the more popular hydrophobicity scales are shown in Table 1.5. According to Eq. 1.4, the more negative the values are in Table 1.5, the more hydrophobic the amino acid will be. There are discrepancies between the scales, either in the absolute values or in their relative order, which can be attributed to empirical details of each experimental approach. For example, the most hydrophobic amino acid in the GES scale (Engelman et al. 1986) is F, whereas in the Kyte–Doolitle (1982) scale it is I.

Table 1.5 Hydrophobicity scales for partitioning of the amino acids into non polar phases: (I) (Kyte and Doolittle 1982), (II) (Eisenberg et al. 1984), (III) Engelman et al. (GES) (1986), (IV) Wimley et al. (1996), (V) Wimley and White (1996). Values of $\Delta_{transf}G$ (kJ mol^{-1})

Amino acid			(I)	(II)	(III)	(IV)	(V)	
Isoleucine	I	Aliphatic	−18.83	−5.77	−12.97	−54.27	−1.30	
Phenylalanine	F	Aromatic	−11.72	−4.98	−15.48	−64.77	−4.73	More hydrophobic
Valine	V	Aliphatic	−17.57	−4.52	−10.88	−45.52	0.29	
Leucine	L	Aliphatic	−15.90	−4.44	−11.72	−49.02	−2.34	
Tryptophan	W	Aromatic	3.77	−3.39	−7.95	−33.26	−7.74	
Methionine	M	Aliphatic	−7.95	−2.68	−14.23	−59.52	−0.96	
Alanine	A	Aliphatic	−7.53	−2.59	6.69	28.01	0.71	
Glycine	G	Aliphatic	1.67	−2.01	−4.18	−17.51	0.04	
Cysteine	C	Aliphatic	−10.46	−1.21	−8.37	−35.01	−1.00	
Tyrosine	Y	Aromatic	5.44	−1.09	2.93	12.25	−3.93	
Proline	P	Aliphatic	6.69	−0.50	0.84	3.50	1.88	
Threonine	T	Aliphatic	2.93	0.21	−5.02	−21.01	0.59	
Serine	S	Aliphatic	3.35	0.75	−2.51	−10.50	0.54	
Histidine	H	Basic	13.39	1.67	12.55	52.52	4.02	
Glutamic acid	E	Acidic	14.64	3.10	34.31	142.55	8.45	
Asparagine	N	Amide	14.64	3.26	20.08	84.03	1.76	
Glutamine	Q	Amide	14.64	3.56	17.15	71.77	2.43	More hydrophilic
Aspartic acid	D	Acidic	14.64	3.77	38.49	161.05	5.15	
Lysine	K	Basic	16.32	6.28	36.82	154.05	4.14	
Arginine	R	Basic	18.83	10.59	51.46	215.32	3.39	

The Kyte-Doolittle scale (I) is based on data obtained for the partitioning from other authors. In turn, the scale developed by Eisenberg et al. (II), known as the consensus scale, is based on the calculation of the hydrophobic dipole moments, (an analogue of the electric dipole moment of a cluster of charges) (Eisenberg et al. 1982) of defined polypeptide chains and the calculation of the Gibbs energy required for the partitioning. The Engelman or GES scale (III) is based on the determination of the Gibbs energy involved in the partitioning of α-helices between water and membranes. The Wimley-White scale is based on the use of the pentapeptide Ace-WLxLL, x being any of the 20 natural amino acids. Actually, two scales have been developed by these latter authors, one using n-octanol as the apolar phase (IV) (Wimley et al. 1996), and the other using bilayers of 1-palmitoyl-2-oleoyl-*sn*-glycero-3-phosphocholine (POPC) (V) (Wimley and White 1996). Interestingly, they have demonstrated that there is a strong correlation between both scales. In principle, this should be attributed to the fact that water-saturated n-octanol presents a micellar structure which seems to mimic the phospholipid molecular arrangement (Franks et al. 1993; White 2007).

IMPs hydrophobicity scales that display the distribution of hydrophobic segments in a linear peptide sequence can be used for the prediction of the two-dimensional topology of a protein. For this purpose, the average Gibbs energy of transfer for a defined number of amino acids (window) is plotted as a function of the first position. Kyte and Doolitle were the first to introduce the hydropathy plot analysis to classify chains in proteins structure. The window is then moved one position across the sequence each time and every set receives a hydrophobicity value, which is plotted to obtain the whole hydrophobicity plot for the protein. The representation will show well-defined maxima which can be assigned to the membrane spanning regions (Fig. 1.15). The outcome is strongly dependent on the hydrophobic scale selected, as well as the size of the window. The latter may range between 5 and 25 amino acids, depending on the specific case, but for bilayer spanning α-helices, a typical value is 18. It should be remarked here that hydrophobicity plots constructed by using the so-called combined scale of Wimley and White, which consists of subtracting the values of the water-POPC interface from the values of the water octanol scale, show favorable peaks that correspond to known transmembrane helices.

Although until recently it was thought that the hydrophobic effect was the most important contribution to protein folding, the combined effect of ionic interactions and hydrogen bonds between residues located in different α-helices of the same protein seem to contribute to a similar extent. Ionic interactions have a dual effect, because sometimes they can be relevant not only for achieving the folded state, but also may be involved in specific changes related with the activity of the protein. This is the case in LacY, where a number of residues involved in the molecular mechanism of the protein also participate in ionic interactions that stabilize its native state. The ionizable residues present within the bilayer are neutralized in order to stabilize the folded structure by forming ion pairs or hydrogen bonds. For

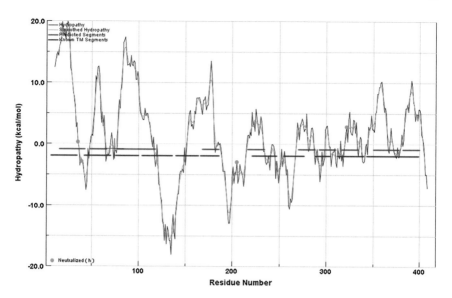

Fig. 1.15 Hydrophopic plot for LacY obtained by using the MPEs (http://blanco.biomol.uci.edu/hydrophobicity_scales.html) software for getting the prediction of TM segments and the ocatnol-water scale

Fig. 1.16 Helical wheel representation of the transmembrane portion of domain VII in Lactose permease of *E. coli*. Reprinted with permission from Voss et al. (1997). Reprinted with permission Biochemistry. American Chemical Society

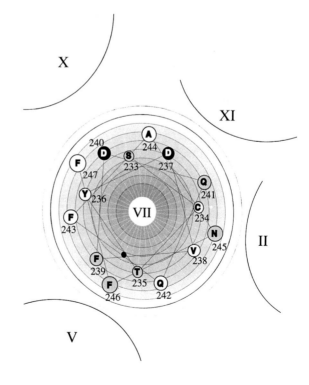

instance, in LacY, ion pairs are formed between E325 (in helix X) and R302 (in helix IX), D240 (in loop VII/VIII) and K319 (in helix X), while R144 (in helix V) forms a salt brige with E126 (in helix IV), and D237 interacts (in loop VII/VIII) with K358 (helix XI). In Fig. 1.16, a helical wheel representation of a portion of the transmembrane part of helix VII of LacY is shown, highlighting some of the charged residues.

Some of these residues are involved in the transport function of this protein. This is not the only case where ionizable residues are found in transmembrane α-helices. Another well-known example is residue K216 in bacterial rhodopsin, which forms a Schiff base with the retinal prosthetic group.

As illustrated in Fig. 1.16, the wheel representation shows the amphipatic nature of the α-helix, with many hydrophobic residues concentrated on one side of the helix (left in the Figure) and relatively more polar residues on the other side (top-right in the Fig. 1.11).

1.4 Micro- and Nanostructure of Biomembranes

The original F-MMM Model provided an insight at the nanoscale level of biomembrane structure. The model was depicted as a mosaic composed of different kinds of IMPs, peripheral proteins, and glycoproteins, which were embedded in phospholipids, glycolipids, and cholesterol (Singer and Nicolson 1972). The complex mixture of lipid and protein components was described as a two-dimensional fluid structure where lipids and proteins were in constant motion. Actually the dynamic nature of the membrane was one of the hints of the F-MMM. There are two main modes of motion: rotational (around the axis of the lipid or the protein) and translational (along the plane defined by the bilayer). A third mode of motion, the transverse diffusion across the perpendicular axis of the bilayer, is less likely to occur because it is energetically unfavorable due to the hydrophobic nature of the bilayer core that is to be traversed. For phospholipid molecules this is an event that may occur spontaneously once every several hours, in absence of specific MPs named flippases, transverse diffusion or flip-flop for proteins is infrequent but may occur. The F-MMM has evolved over time after novel experimental concepts that have recently been revisited (Goñi 2014) pointing to different concepts that should conform with the F-MMM to maintain its validity. These new insides include a better understanding of the high density of transmembrane proteins; proteins that bind transiently at the membrane surface; the existence of phases different from the lamellar phase and their possible physiological relevance; the curvature of the membrane which depends on the geometry and nanomechanical properties of lipids and proteins (see Chap. 4); the lateral heterogeneity of membranes caused by non-ideal mixing; and the physicochemical properties of the membrane components or deviations from the equilibrium due to transbilayer lipid diffusion which may occur under specific conditions. Let's review some of these aspects in more detail to illustrate how biophysical research on lipids and proteins introduced progresses in our present view of membrane structure at the nanoscale level.

At the time that the F-MMM was proposed, it was already envisioned that several mechanisms might potentially reduce or restrict the lateral diffusion of membrane components. Most of these mechanisms, including the lateral self-segregation of glycoproteins in the membrane plane, the formation of enriched protein and lipid domains, the interaction of glycoproteins with extracellular components, and peripheral membrane protein interactions or cytoskeletal interactions, have been now experimentally demonstrated. An illustrative example is the reduction of the translational diffusion of IMPs evidenced experimentally by using single-particle tracking (SPT), fluorescence correlation spectroscopy (FCS), and by fluorescence recovery after photobleaching (FRAP) techniques. This last method consists of a labeling the membrane with a specific fluorescent probe (lipid or protein) and the monitoring of light emitted through a fluorescence microscope. A short laser pulse is applied to a small region of the membrane, which causes photobleaching (destruction) of the fluorescent probe with the consequent fading of the emission light in the region of the pulse. The fluorescence is gradually recovered within a short period of time when intact probes from non-bleached areas diffuse into the selected spot. The mean square distance moved by the label from its origin ($t = 0$) in a interval of time (Δt) is expressed by the equation

$$x = (4D\Delta t)^{1/2} \qquad (1.5)$$

where D is the diffusion coefficient of the label. Although the technique was initially intended only for lipids, it can be also applied to measure D values of IMPs by using green fluorescent protein (GFP) fusion proteins. While D values for lipids can be ~1 μm^2 s^{-1}, values for proteins range from extremely mobile proteins such as rhodopsin which has an estimated D of 0.4 μm^2 s^{-1} to less mobile proteins like the peripheral glycoprotein fibronectin which shows D values less than 10^{-4} μm^2 s^{-1}. These are examples that have modified the idea of free in-plane diffusion as conceived in the original F-MMM. Actually, the low mobility of fibronectin is due, as in other cases, to anchorage to actin filaments through specific IMPs that link the extracellular matrix to the cytoskeleton.

We have emphasized the basic importance of the hydrophobic interactions in determining the basic microstructure of cell membrane. However, the idea that membranes are formed by homogeneous regions of phospholipid bilayers which behave as a matrix through which proteins can diffuse is far from reality. The existence of lipid domains is known from classical studies on the lipid mixing properties of phospholipids (Shimshick and McConnell 1973). These domains are characterized by a specific chemical composition at a given temperature and provide evidence for the possible coexistence of different fluid phases delimited by a physical boundary within the membrane.

Membranes and the lipids from which they are made are mesomorphic which means that they present intermediate properties between solid and liquid (Fig. 1.17).

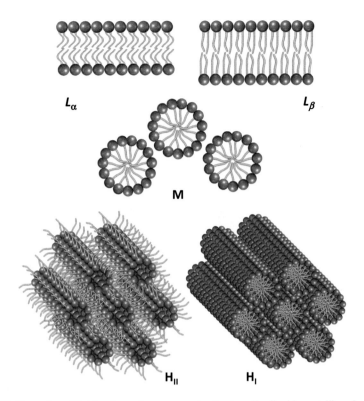

Fig. 1.17 Examples of lipidic phases in excess water. L_α, lamellar liquid crystalline; L_β, lamellar gel; M, micellar; and H_I (tubular) and H_{II}, (inverted), hexagonal phases

Upon varying the water content or the temperature or both, lipids have the ability to adopt several mesophases of which the biological implications are not totally clear. Most phospholipids when fully hydrated at a certain temperature form a lamellar phase, called the solid-like gel state (L_β), and upon heating, the acyl chains melt to a liquid-crystalline state (called L_α), which is often referred as the fluid state. The transition between L_α and L_β occurs at a defined temperature (T_m) which is characteristic of each phospholipid species and mixture. In the case of eukaryotic cell membranes made up of cholesterol-sphingomyelin and unsaturated lipid mixtures one can make additional distinctions between liquid-ordered (L_o) and liquid-disordered (L_d) phases (Mouritsen 2011). The possibility that biological membranes do not have to be in the in L_d phase continuously and that some domains could persist in the L_o phase is consistent with the existence of lipid rafts (see Sect. 3.4). These could occur in membrane regions enriched in cholesterol, sphingolipids and membrane proteins. Many GPI-anchored proteins or IMPs with double acylation and/or long hydrophobic chains have been suggested be in such rafts.

In fluid phases, lipids can adopt other macromolecular arrangements such as globular micelles (M) or normal tubular (H_I) and inverted hexagonal (H_{II}) or cubic

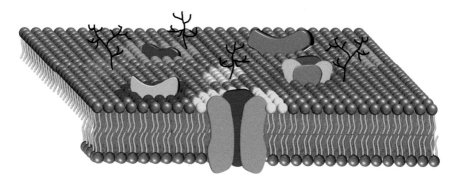

Fig. 1.18 Illustation of the modified fluid-mosaic membrane model based on Escribá et al. (2008)

(Q_{II}) phases (Fig. 1.17). Whilst in H_I phase the polar head is found on the outside of the structure and the fatty acyl chains point toward the center, in H_{II} cylinders, lipid polar headgroups point to the core and the fluid acyl chains point to the hydrophobic region. In both cases, a cross-section of the structures exhibits a hexagonal lattice. Q_{II} phases are thermodynamically stable and consist of a curved bicontinuous lipid bilayer in three dimensions, separating two congruent networks of water channels. There are three common types of bicontinuous cubic phases with different space group symmetries. As will be further discussed in Sect. 2.4.6 and Chap. 4 the particular disposition of a phospholipid for one of these macromolecular organizations depends on the shape of the molecule and, physically, form its intrinsic curvature (Sects. 2.5 and 4.6).

Many phospholipids in natural membranes have a strong tendency to form nonlamellar phases when dispersed in water, which introduces the question of their existence, even transiently, in natural membranes. Such transient non-lamellar organization behavior is thought to be relevant in several membrane processes such as fusion, exocytosis, interbilayer tight junctions, ion permeability or the insertion of IMPs in the membrane.

Nicolson has himself (Nicolson 2014) integrated most of the new concepts and evidence into a model which greatly improves our understanding of the membrane structure at the nanoscale level, inspired by the original F-MMM. As can be seen in Fig. 1.18 (Escribá et al. 2008) the updated model incorporates recent information on membrane domains, lipid rafts and cytoskeletal fencing.

References

Bogdanov M, Sun J, Kaback HR, Dowhan W. A phospholipid acts as a chaperone in assembly of a membrane transport protein. Journal of Biological Chemistry. 1996;271(20):11615–8.
Cymer F, von Heijne G, White SH. Mechanisms of Integral Membrane Protein Insertion and Folding. Journal of molecular biology. 2014;427(5):999–1022.
Danielli JF, Davson H. A contribution to the theory of permeability of thin films. Journal of cellular and comparative physiology. 1935;5(4):495–508.

Dowhan W, Bogdanov M. Lipid-dependent membrane protein topogenesis. Annual review of biochemistry. 2009;78:515–40.

Eisenberg D, Schwarz E, Komaromy M, Wall R. Analysis of membrane and surface protein sequences with the hydrophobic moment plot. Journal of molecular biology. 1984;179(1):125–42.

Eisenberg D, Weiss RM, Terwilliger TC, Wilcox W. Hydrophobic moments and protein structure. Faraday Symposia of the Chemical Society. 1982;17:109.

Engelman DM, Steitz TA, Goldman A. Identifying nonpolar transbilayer helices in amino acid sequences of membrane proteins. Annual review of biophysics and biophysical chemistry. 1986;15:321–53.

Escribá PV, González-Ros JM, Goñi FM, Kinnunen PKJ, Vigh L, Sánchez-Magraner L, et al. Membranes: A meeting point for lipids, proteins and therapies: Translational Medicine. Journal of cellular and molecular medicine. 2008;12(3):829–75.

Franks NP, Abraham MH, Lieb WR. Molecular organization of liquid n-octanol: An X-ray diffraction analysis. Journal of pharmaceutical sciences. 1993;82:466–70.

Goñi FM. The basic structure and dynamics of cell membranes: an update of the Singer-Nicolson model. Biochimica et Biophysica Acta (BBA)-Biomembranes. 2014;1838(6):1467–76.

Gorter E, Grendel F. on Bimolecular Layers of Lipoids on the Chromocytes of the Blood. The Journal of experimental medicine. 1925;41(4):439–43.

Kyte J, Doolittle RF. A simple method for displaying the hydropathic character of a protein. Journal of molecular biology. 1982;157:105–32.

Martz E. Protein Explorer: Easy yet powerful macromolecular visualization. Trends in biochemical sciences. 2002;27(2):107–9.

Merino-Montero S, Montero MT, Hernández-Borrell J. Effects of lactose permease of *Escherichia coli* on the anisotropy and electrostatic surface potential of liposomes. Biophysical chemistry. 2006;119(1):101–5.

Mouritsen OG. Model answers to lipid membrane questions. Cold Spring Harbor perspectives in biology. 2011. p. 1–15.

Nicolson GL. The Fluid-Mosaic Model of Membrane Structure: still relevant to understanding the structure, function and dynamics of biological membranes after more than 40 years. Biochimica et Biophysica Acta (BBA)-Biomembranes. 2014;1838(6):1451–66.

Nozaki Y, Tanford C. The solubility of amino acids and two glycine peptides in aqueous ethanol and dioxane solutions. Establishment of a hydrophobicity scale. Journal of Biological Chemistry. 1971;246(7):2211–7.

Picas L, Milhiet PE, Hernández-Borrell J. Atomic force microscopy: A versatile tool to probe the physical and chemical properties of supported membranes at the nanoscale. Chemistry and physics of lipids. 2012. p. 845–60.

Ramachandran GN, Ramakrishnan C, Sasisekharan V. Stereochemistry of polypeptide chain configurations. J. Mol. Biol. 1963;7(1):95–9.

Shimshick EJ, McConnell HM. Lateral phase separation in phospholipid membranes. Biochemistry. 1973;12(12):2351–60.

Singer SJ, Nicolson GL. The fluid mosaic model of the structure of cell membranes. Science. 1972;175(23):720–31.

Sun J, Frillingos S, Kaback HR. Binding of monoclonal antibody 4B1 to homologs of the lactose permease of *Escherichia coli*. Protein science. 1997;6(7):1503–10.

Voss J, Hubbell WL, Hernandez-Borrell J, Kaback HR. Site-directed spin-labeling of transmembrane domain VII and the 4B1 antibody epitope in the lactose permease of *Escherichia coli*. Biochemistry. 1997;36(49):15055–61.

Wang X, Bogdanov M, Dowhan W. Topology of polytopic membrane protein subdomains is dictated by membrane phospholipid composition. The EMBO journal. 2002;21(21):5673–81.

White SH. Membrane protein insertion: the biology-physics nexus. The Journal of general physiology. 2007;129(5):363–9.

White SH, Wimley WC, Ladokhin AS, Hristova K. Protein folding in membranes: Determining energetics of peptide-bilayer interactions. Methods in enzymology. 1998;295:62–87.

Wimley WC, Creamer TP, White SH. Solvation energies of amino acid side chains and backbone in a family of host-guest pentapeptides. Biochemistry. 1996;35:5109–24.

Wimley WC, White SH. Experimentally determined hydrophobicity scale for proteins at membrane interfaces. Nature Structural & Molecular Biology. 1996;3(10):842–8.

Chapter 2
Physicochemical Properties of Lipids and Macromolecules in Higher Level Organization

Abstract This chapter relates in a very concise way, how the physicochemical properties of membrane lipids determine the formation of self-segregated structures. The most common model methods used to understand the influence of lipid organization in membranes, lipid monolayers, liposomes and supported lipid bilayers, are reviewed as well for their suitability in the investigation of lipid-membrane protein interactions.

Keywords Langmuir-Blodgett films · Surface phases · Hydrophobic effect · Lipid polymorphism · Liposomes · Guvs · Transition temperature · Fluorescence anisotropy · ^{31}P-NMR

2.1 Lipid Monolayers at the Interface: Two Dimensional Structures

Monolayers formed at the air-water interface are of great interest as earlier exemplified by the work of Gorter and Grendel (see references of Chap. 1) that led to the proposal of a bilayer model for biological membranes. Their insightful and impactful conclusions were, remarkably, based on conventional compression experiments of lipids extracted from red blood cells at the lipid-water interface. Monolayers represent only half of a bilayer but are useful since lateral pressure and physicochemical conditions can be controlled experimentally. Indeed, monolayers have been extensively used to investigate the interaction between proteins, peptides and drugs with lipids and lipid mixtures by mimicking the lipid-water interface of biological membranes.

2.1.1 Phases at the Air Water Interface

Monolayers are formed by the deposition of lipids dissolved in a volatile solvent (i.e. Cl_2CH_2, Cl_3CH or CH_3OH) at the air-water interface. After a period of solvent evaporation this leads to the spontaneous formation of monomolecular lipid films. Aqueous suspensions and dispersions of lipids such as liposomes will also spontaneously form monolayers at the air-water interface, and the properties of those monolayers provide useful information for understanding the properties of the monolayers and related lipid structures such as liposomes (see, e.g., lung surfactant, below). Conventional experimental set-ups have been derived from pioneering observations that a film could be created in a basic film balance instrument consisting of a cuvette or trough able to contain a subphase where the film was confined between two barriers, one mobile on one end of the trough, and the other floating at the other end. The force exerted by the film on the floating barrier was directly measured by means of a sensitive balance. The designs that are available in modern instruments are based on this earlier approach. Modern troughs are available in different volumes and sizes with one or two mobile barriers made of Teflon to avoid any leakage past the limits of the working surface. When the lipid film is compressed between the two barriers a measure of the *surface pressure* (π) versus the molecular area (A) is obtained. This is normally called a compression isotherm and it is characteristic of the phospholipid (or phospholipid mixture) spread at the air-water interface. π is defined as the difference between the *surface tension* (γ) of clean surface (water) and that with the monolayer,

$$\pi = \gamma^w - \gamma^m \tag{2.1}$$

Upon lateral compression of a film, the molecular area, referring to the area occupied by a single molecule, decreases in response to the increase in the lateral pressure exerted by the barriers (Fig. 2.1b, c), up to a value of pressure termed the collapse pressure (Fig. 2.1d) where the molecules cannot be further compressed without losing the unimolecular arrangement of the monolayer. At this point during compression both one and two-dimensional phases of the lipid will usually be formed. When monolayers are compressed very slowly, the collapse pressure is usually the same as the equilibrium pressure achieved when lipid suspensions and monolayers coexist over a longer period of time. When monolayers are compressed rapidly, they might enter a metastable one or two-dimensional state, that can relax to the equilibrium pressure over a time that is highly dependent on the composition and the rate of compression (such is the case in lung surfactant).

When kept below their corresponding critical temperatures and surface pressure, phospholipid monolayers can be compressed and may undergo several phase transitions. Four main monolayer phases in two dimensions have been described: the gaseous (G), liquid-expanded (L_e), liquid-condensed (L_c) and solid (S) phases. Monolayers in the he L_e phase have physicochemical properties similar to those expected of the 3-dimensional L_α phase (see Sect. 2.6.1). A high degree of care is necessary to obtain an accurate π determination during the operation of these

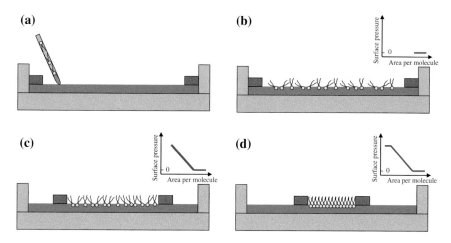

Fig. 2.1 Cartoon showing the compression with a two floating mobile barriers of a lipid monolayer deposited at the air water-interface and the shapes of the compression isotherms obtained. The surface pressure can be determined by either measuring the force exerted by the film on one of the barriers (original Langmuir method) or by measuring the surface tension in the interfacial film compared to that in the clean solvent (the Lagmuir-Wilhelmy method)

experiments. In biomembrane research, the most common phospholipid used as a reference for isotherm properties is DPPC. This saturated homoacid phospholipid is one of the main constituents of lung surfactant and its compression isotherm (Fig. 2.2) presents very defined and well-established features that can be used as a reference to compare results between laboratories.

At low surface pressures, molecules occupy large areas in such a way that the monolayers can be expanded indefinitely without incurring a phase change. In comparison to what happens with matter in three dimensions, wherein molecules range over large volumes at low pressures, the monolayer is considered to be in G state. After the lateral compression of films of phospholipids in the G phase a L_e phase appears which is characterized by a decrease in intermolecular distance. Further compression leads to the appearance of an L_c phase followed by a region of $L_e - L_c$ phase coexistence, usually characterized by a plateau in the isotherm. L_c is characterized by maximum packing and a minimum molecular area at the interface. When the lipids have been fully converted to the L_c phase the surface pressure rises steeply. Further compression of a fully converted L_c phase, will result in hte collapse of the monolayer into three-dimensional forms without further increases in surface pressure. One direct way to visualize the events during compression is the use of fluorescent phospholipid labels like 1-palmitoyl-2-[12-{(7-nitro-2–1,3-benzoxadiaol-4-yl)amino}]phosphatidylcholine (NBD-PC) which inserts preferentially into L_e phase monolayers. The dark domains in Fig. 2.2a–d represent the L_c phase. As the compression progresses the domains increase in size (after Nag and Keough 1993).

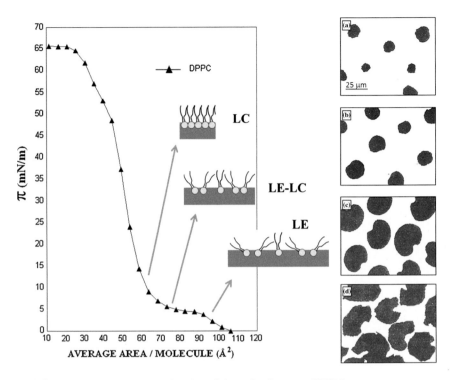

Fig. 2.2 Surface pressure (π) as a function of the molecular area of DPPC at room temperature. Cartoons indicate the theoretical disposition at the air-water interface of the molecule. Typical epifluorescence images of the DPPC:NBD-PC (99:1, mol/mol) monolayer at 20 °C at surface pressures of: 4 (**a**), 7 (**b**), 10 (**c**) and 15 (**d**) mN m^{-1}

2.1.2 Monolayer Compressibility

The *compressibility* (C_s) of a monolayer is defined by

$$C_s = -\frac{1}{A}\left(\frac{\partial A}{\partial \pi}\right)_{T,n} \tag{2.2}$$

where A is the molecular area of the phospholipid. It can be calculated from the slope of the curve of π versus the   molecular area. Often it is expressed as the reciprocal of the *compressional modulus* C_s^{-1} that is

$$C_s^{-1} = (-A) \cdot \left(\frac{\partial \pi}{\partial A}\right)_{T,n} \tag{2.3}$$

In practical terms, the derivative of the experimental data is computed by fitting a straight line to a window of an area width of ~0.2 nm^2 mol^{-1} around any given surface pressure value, so that experimental noise is filtered out. It follows that the C_s^{-1} value for the clean air-water interface is zero, with an equivalent value being found experimentally for monolayers in G state; its value ranges between 5 and 50 mN m^{-1} for the L_e state and between 100 and 250 mN m^{-1} for the L_c state. Solid state compressional moduli may range from 1000 up to 2000 mN m^{-1} (Davies 1963).

2.1.3 Mixing Properties of the Monolayers at the Interface

The determination of mixing properties of lipids in monolayers can help in understanding their mixing properties in bilayers or membranes. Most useful in this regard is determining the magnitudes of the thermodynamic characteristics associated with the process. The molecular area of an ideally mixed monolayer of two components can be calculated according to

$$A^{id} = \chi_1 A_1 + \chi_2 A_2 \tag{2.4}$$

where A^{id} is the monolayer and χ_1, A_1 and χ_2, A_2 are the molar fractions and the molecular areas of the pure components 1 and 2, respectively. The excess area, A^E, for a binary monolayer can be expressed as follows

$$A^E = A_{12} - (\chi_1 A_1 + \chi_2 A_2) \tag{2.5}$$

where A_{12} is the molecular area of the mixed monolayer. Negative values of A^E indicate attractive forces between molecules while positive values indicate repulsive forces. The interaction between two phospholipid components in a mixed monolayer, at a constant π and temperature, can be evaluated by calculating the excess *Gibbs energy* (G^E), which is given by

$$G^E = \int_0^\pi [A_{12} - (\chi_1 A_1 + \chi_2 A_2)] d\pi \tag{2.6}$$

$G^E < 0$ means that the mixing between both components is favored and, conversely $G^E > 0$ indicates that mixing of molecules is disfavored. The meaning of G^E can be better understood by defining the *Gibbs energy of mixing* as follows

$$\Delta_{mix} G = \Delta_{mix} G^{id} + G^E \tag{2.7}$$

where the first term, the ideal Gibbs energy of mixing ($\Delta_{mix} G^{id}$), can be calculated from the equation

$$\Delta_{mix} G^{id} = RT(\chi_1 Ln\chi_1 + \chi_2 Ln\chi_2) \tag{2.8}$$

where R is the universal gas constant and T is the temperature. We find that G^E at a defined temperature depends on the magnitude of available experimental conditions of composition and intrinsic molecular areas. All these thermodynamic magnitudes will vary when there are interactions between peptides or proteins in the monolayer.

2.2 Langmuir-Blodgett Films

Langmuir-Blodgett (LB) techniques (Petty 1996) were developed in the first decades of the twentieth century. The strategy consisted of raising a solid substrate through a monolayer that was formed at the interface as is depicted in diagrams in Fig. 2.3a–c. If the substrate is hydrophilic the headgroups become attached onto the substrate with the acyl chains facing the air as it passes upward (upstroke). Conversely, by using a hydrophobic substrate and passing it downward (downstroke) through the phospholipid monolayer, the acyl chains will be attached onto the solid surface and the headgroups will face the water subphase. The technique has the advantage that ionic strength, pH, temperature and surface pressure can be kept constant in such a way that the monolayer is theoretically transferred onto the substrate with the same structural properties that it had at the air-water interface. The transfer ratio (TR) value gives information about the yield of coverage of the solid substrate, it being nearly 1 for a fully effective transfer. In this way a lipid monolayer can be transferred at any desired point of the isotherm (Fig. 2.2) and the features of the different states (i.e. L_e or L_c) can be scrutinized by using suitable surface imaging techniques (i.e. XPS or AFM). A point of criticism may arise from the possible difference between the lateral surface pressure at the air-water interface and the lateral pressure in the transferred LB film. Although, many mechanical aspects such as the lifting speed have been improved over the years, it is true that the transferred LB film might be a little more relaxed, i.e., at a slightly lower lateral pressure or packing density than the film from which it was transferred.

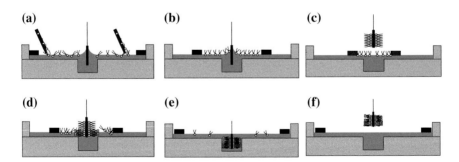

Fig. 2.3 Langmuir-Blodgett extraction: lipids are spread at the air-water interface (**a**) and compressed to the desired surface pressure (**b**). A first monolayer is extracted by lifting up a substrate through the interface (**c**), followed by passing the substrate downward into the subphase to attach a second monolayer (**d**, **e**). Subsequent rising of the substrate (**f**) produces a substrate that is coated with a double monolayer or bilayer

The LB technique can be further exploited to obtain bilayers by double monolayer deposition. The procedure may be performed as follows: the lipids of interest are spread at the air-water interface and at the convenient surface pressure transferred onto a solid hydrophilic support. Afterwards, a second monolayer is transferred onto the one deposited before on the down-stroke to obtain a supported bilayer Fig. 2.3c–e). The technique can be used to obtain asymmetric bilayers with different surface pressure at the apical and distal monolayers of the bilayer, with the same or different composition.

2.3 Structures at the Air-Water Interface

Several direct optical methods applied at the air-water interface or indirect methods applied to the extracted LB films, can be used to investigate the structure of the lipid films. The first direct visual evidence of a L_e-L_c phase equilibrium was provided by fluorescence microscopy of monolayers (FM) (Peters and Beck 1983) which is based on the different affinity of fluorescent phospholipid analogs for the different phases. A drawback of FM is that the presence of fluorescent phospholipid analogs used as reporters induced some distortions in the packing structure of the phases, so that the concentration of probe in the monolayer (or bilayer) needs to be kept as low as practically possible. Different techniques for studying monolayers without the use of probe molecules have become available. These include Brewster angle microscopy (BAM) (Hoenig and Moebius 1991), atomic force microscopy (AFM), X-ray reflectivity and ellipsometry, and surface Raman spectroscopy. It is worth noting that distortions and artefactual behavior can easily occur in all monolayer measurements, and great care is necessary in both experimental execution and the interpretation of results.

2.3.1 Brewster Angle Microscopy

BAM is a microscopic method based upon the absence of reflection of p-polarized light from a clean interface at a certain angle (Brewster angle). When lipids are incorporated into the air-water interface the refractive index varies and the covered part of the surface reflects light at the Brewster angle for the clean surface. The different monolayer phases have different reflectances so BAM is capable of providing information on the L_e-L_c transitions in lipid monolayers along with information about domain morphology and the possible order and local packing of the domains. As an example the surface pressure-area (π-A) isotherm of a pure POPE monolayer at 24.0 ± 0.2 °C is shown in Fig. 2.4 along with BAM images taken at different surface pressures (Domènech et al. 2007a). At very low surface pressures the monolayer is in the G state (Fig. 2.4a). While the isotherm of POPE is in the L_e phase, BAM images reveal the existence of bright spots (Fig. 2.4b). Their number does not appear to increase above $\pi \sim 6$ mN m^{-1}, although they

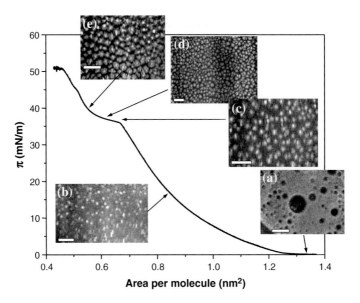

Fig. 2.4 Compression isotherm of POPE showing the typical L_e-L_c coexistence plateau that extends from ~0.68 to ~0.57 $nm^2 mol^{-1}$ and the collapse at 50.7 $mN\ m^{-1}$. BAM images were captured at 0.5 (**a**), 15 (**b**), 36 (**c**), 38 (**d**) and 39 (**e**) $mN\ m^{-1.}$ Adapted with permission from Domènech et al. (2007a) © 1997. American Chemical Society

become increasingly brighter upon compression. Images do not reveal any new emerging structures until the plateau region when the growth of globular domains with increased BAM reflectivity is observed (Fig. 2.4c, d) which is consistent with a L_e-L_c phase transition. Bright spots present at lower pressures act as nucleation points of the new domains, whose structure is typical of condense L_c domains in phospholipid monolayers. The major drawback of BAM is the limited resolution, which falls within the range of micrometers. In this regard, it could be of interest to compare BAM with AFM images of the same monolayer.

2.3.2 Atomic-Force Microscopy (AFM)

AFM is a near-field microcopy technique forming part of the wider group of scanning probe microscopy techniques (SPMs), which were developed after the invention of scanning tunneling microcopy (STM) by Binnig et al. (1982). AFM provides outstanding vertical and lateral resolution (better than 0.1 nm and 1 nm, respectively). AFM scanning of POPE monolayers complements the previous BAM observations. The POPE isotherm is presented in Fig. 2.5 along with AFM images of LBs extracted at different surface pressures.

The LB extracted at 10 $mN\ m^{-1}$ (Fig. 2.5a) is featureless and corresponds to the L_e phase. At 30 $mN\ m^{-1}$ numerous randomly distributed vacancies with

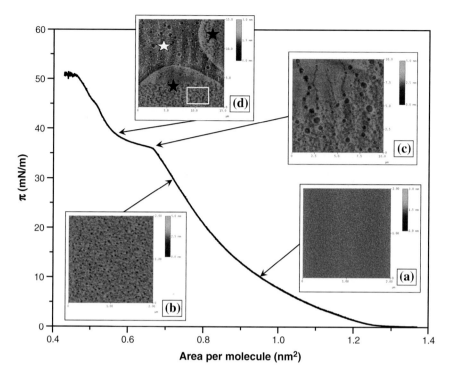

Fig. 2.5 Compression isotherm of POPE showing the typical L_e-L_c coexistence plateau that extends from ~0.68 to ~0.57 nm^2 mol^{-1} and the collapses at 50.7 mN m^{-1}. AFM images were acquired at 10 (**a**), 30 (**b**), 36 (**c**) and 39 (**d**) mN m^{-1} Adapted with permission from Domènech et al. (2007a) © 1997. American Chemical Society

a diameter of ~75 nm are observed (Fig. 2.5b). As surface pressure increases these vacancies coalesce resulting in regions with diameters ranging from ~90 to ~500 nm at 36 mN m^{-1} (Fig. 2.5c). When the pressure reaches 39 mN m^{-1}, these vacancies have grown into much more complex structures (Fig. 2.5d). This last image shows two well defined domains with a step-height difference of ~0.5 nm that represents the typical step height between L_c and L_e phases. The thinner domain (white star) represents the L_e and the thicker domain (black star) represents the L_c phase. It is worthwhile mentioning that the LC structures are of a similar size as those observed in Fig. 2.3e.

2.4 Protein- and Peptide-Lipid Interactions in Monolayers

Different experimental approaches can be used to investigate the interaction between proteins and peptides with lipid monolayers. One experimental approach used to determine the nature of the interaction has been to inject the molecule of interest into the subphase beneath the lipid monolayer. Molecules that have some hydrophobic components such as peptides or proteins that are dissolved in aqueous solutions, will display some tendency to migrate into the air-water interface. They will "compete" for surface space with lipids in very loosely packed monolayers. Subsequent compression of the mixed monolayers can yield useful information about the thermodynamics of the mixed lipid-protein/peptide film. If there is insertion of the protein or peptide into the lipid monolayer there will be changes in the isotherms obtained by compression of the film. Usually, moderate expansions to larger nominal lipid molecular areas are observed most likely due to the intercalation of the peptide or protein between the lipid molecules. In some instances, the insertion leads to a condensation of the film to lower nominal molecular areas per lipid. The repulsive or atractive manner of the interaction can be quantitatively evaluated within the surface thermodynamics framework described in Sect. 2.1.3. Protein or peptide molecules do not need to insert into the film to have an influence on the lipid packing. Some molecules may interact with or attach to the headgroup region of the phospholipids causing effects on the isotherms. In some cases, especially when the protein or peptide is itself highly amphipathic, it can cause desorption of the lipid monolayer into the subphase by the formation of micelles.

Another simple and common method used in peptide research consists of following the adsorption of the molecules of interest into the air-water interface containing a lipid monolayer at constant area and recording the changes in the surface pressure as a function of the time. The interaction can be also measured, as will be discussed in Sect. 2.4.4, by keeping the surface pressure constant while measuring the increase or decrease of the area covered by the monolayer as an indicator of the interaction (Maget-Dana 1999).

2.4.1 Interfacial Studies for Understanding Enzyme Activity

Lipid monolayers are used as model membranes. They can also be exploited to investigate the action of soluble enzymes such as phospholipases (PLs), on specific phospholipids, a process which was pioneered by the De Haas Group in the 60s and 70s (Verger et al. 1973). PLs hydrolyze phospholipids at the membrane interface liberating fatty acids and for this reason the monolayer is a well suited membrane model for the investigation of the molecular mechanism of action of PLs.

Members of the extracellular secretory phospholipase A_2 family (PLsA$_2$) are the most extensively studied enzymes that catalyze reactions at the lipid-water interface. PLsA$_2$ catalyze the hydrolysis of the *sn*-2 ester bond of glycerophospholipids leading to free fatty acids and lysophospholipids. PLsA$_2$ undergo a significant increase in their catalytic activity when bound to the surface of phospholipid membranes, a process highly sensitive to some membrane physical properties such as phase, surface charge or curvature. A model for the action of PLsA$_2$ at interfaces was proposed by Verger et al. (Fig. 2.6). This model which basically consisted of two successive equilibria: the first describing the interfacial adsorption and penetration

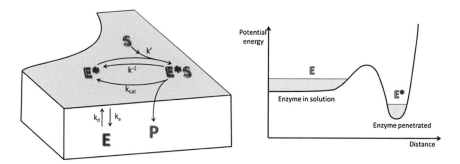

Fig. 2.6   Reaction mechanism of a soluble enzyme with a lipid substrate at an interface

of PLsA$_2$ into the phospholipid monolayer, which leads to an activated form of the enzyme; and the second, describing the formation of the activated enzyme-substrate complex following the Michaelis–Menten model adapted to two dimensions.

2.4.2 Adsorption of Soluble Proteins to Lipid Monolayers

A good example of these experiments is cytochrome c (cyt c), a peripheral membrane protein that is localized at the inner membrane of mitochondria. It is an essential component of the mitochondrial respiratory chain, transferring electrons between the CoQH$_2$-cytochrome c reductase and the cytochrome c oxidase complexes. The mechanism by which cyt c diffuses between the reductant and the oxidant protein substrates may involve translational motion of cyt c in the two-dimensional plane defined by the bilayer, with the protein interacting specifically with cardiolipin (diphosphatidylglycerol).

One can study the interaction and insertion of soluble proteins with lipid monolayers by monitoring the increase in nominal area per lipid molecule at constant surface pressure as a function of time after injecting the protein below the lipid film. In Fig. 2.7, an example of such an experiment where cyt c was injected beneath lipid monolayers of different composition held at constant surface pressure is shown. In these experiments, π was kept at 30 mN m^{-1} while cyt c was injected to reach a final concentration of 2 μM. Changes of surface molecular area (ΔA) with time (t) are measured and experimental data fitted to the a Langmuir-derived isothermal adsorption equation

$$\Delta A = \Delta A_{\text{max}} \frac{(k_i t)^b}{1 + (k_i t)^b} \tag{2.9}$$

where ΔA_{max} is the maximum increase of the lipid molecular area reached in steady-state conditions, k_i is a rate constant determined by the nature of the substances involved and the experimental conditions, and b is a parameter that is

Fig. 2.7 Increase of the surface area A as function of time of monolayers after subsurface injection of cyt c: CL (■), POPE (□), POPC (○), POPE:CL (0.8:0.2, mol/mol) (●), and POPC:CL (0.6:0.4, mol/mol) (▲). Reprinted with permission from Domènech et al. (2007a, b, c). Langmuir. 23 (10):5651–6. © 1997. American Chemical Society

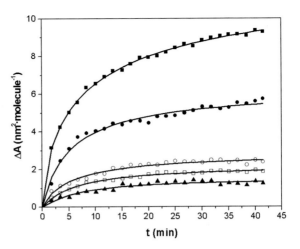

related to the cooperativity of the process. What Fig. 2.7 implies is that while the interaction between cyt c and pure zwitterion phospholipids (POPE and POPC) involves a hydrophobic contribution, the presence of an anionic phospholipid (for instance CL) in the POPE monolayer enhances the penetration of cyt c which can be attributed to specific electrostatic interactions.

2.4.3 Peptide Interaction with Monolayers

An initial recognition between a (soon to be infected) host-cell and a microbial pathogen must take place. This recognition occurs between transmembrane proteins (receptors) present in the membranes of the host-cell and the pathogen, or between pathogen membrane receptors and the outer leaflet moiety of the host-cell membranes. The biophysical study of these interactions using lipid monolayers can shed light into the mechanism of the infection processes. Regrettably, transmembrane proteins are only stabilized in water solution in the presence of surfactant molecules that form micelles or other three-dimensional structures (e.g. bicelles) around the proteins, shielding their hydrophobic moieties. Notably, this handicap has been successfully solved in many cases by a "divide and conquer" strategy consisting of studying only the specific parts of transmembrane proteins (normally peptides that can obtained synthetically) known to be the key elements in the interactions with lipid membranes.

In these types of experiments the selected peptides must themselves possess surface activity. Initially, the saturating concentration of these peptides is set to where no further increase in surface pressure can be obtained. Thereafter any further increase in concentration can be determined by performing surface pressure measurements at increasing peptide concentrations in absence of a lipid monolayer. Next, maximum insertion pressures (MIPs) for a defined lipid monolayer and peptide combination are determined by injecting the peptide underneath the lipid monolayer at different initial pressures (π_i). In these experiments when the peptide-lipid monolayer interaction reaches an equilibrium, a new surface pressure (π_e) is attained. Performing experiments at increasing π_i values yields a representation of the surface pressure increase $\Delta\pi$ ($\Delta\pi = \pi_e - \pi_i$) as a function of π_i. The value for the MIP can be determined from this curve as the intersection point with the x-axis. It is believed that membrane lateral pressure in biological membranes varies from 25 to 30 mN m^{-1} so MIPs values below these numbers will suggest that peptides cannot penetrate the monolayer. Interestingly, the affinity between the lipid monolayer and the peptides can be calculated by adding 1 to the slope of the curve of $\Delta\pi$ as a function of π_i. Negative and positive synergy values represent repulsive and attractive interactions between lipids and peptides, respectively. A zero value represents no interaction.

2.4.4 The Membrane Associated Surfactant Proteins

Lung surfactant is a lipid-protein complex located in the aqueous lining layer of the lung and at the air-water alveolar interface with the principal physiological function of regulation gas exchange activity between the airspaces of the lung and the blood (Pérez-Gil 2008). Lung surfactant also participates in the defense against inhaled pathogens and modulates the function of respiratory cells. It is found in the air spaces in different physical forms with different biophysical activities and with attendant small changes in composition. Pulmonary surfactant is synthesized, assembled and secreted onto the respiratory surface by specialized cells of the alveolar epithelium. It reduces the surface tension of the thin layer of water that covers the lung epithelium. It reduces the work of expansion of the lung surface and it prevents the collapse of the alveoli during expiration. Lack of an operative surfactant system is associated with severe respiratory dysfunction. Lung surfactant has a very particular composition and organization that has been extensively investigated by means of monolayers since it is accepted that the physiological function of lowering surface tension is achieved by a monolayer at the lung air-water interface, possibly associated with adjacent of adhering bilayer structures. More recently techniques that capture the properties of the lung *in vivo* such as the captive bubble method have also been used to study this complex system (Schurch et al. 1992).

Although the composition may vary between species and according to environmental and pathological conditions, lung surfactant is composed of 90 % lipids by weight and 10 % proteins. DPPC is by far the most abundant lipid species in mammalian lung surfactant and usually accounts for 50 % or more of the lipid component. It is essential for producing the very low surface tension (DPPC has a high surface pressure according to Eq. 2.1) observed during lung exhalation which corresponds to surface area reduction and compression of a monolayer at the lung surface. This property is directly related with its saturated acyl chains (16:0), which can adopt a highly packed lateral state (Fig. 2.2). Other species such as PG or PI account for less than 10 % by weight of the total lipid fraction of lung surfactant. The presence of a relatively high PG content in surfactant is unique for this lipid in any mammalian system or membrane. These negatively charged species are thought to participate in selective interactions with cationic hydrophobic proteins. Cholesterol is a significant component, constituting up to about 15 % of the lipid on a molar basis. Lung surfactant also contains four specific proteins, two of them, SP-A and SP-D, are hydrophilic, and the other two, SP-B and SP-C, hydrophobic. The two hydrophobic proteins interact strongly with lipids. Even the two hydrophilic proteins show some propensity to interact with lipids. Their structural models and proposed mode of interaction in the alveolar spaces are presented in Fig. 2.8. SP-B and SP-C participate in the surface activity of the surfactant, while SP-A and SP-D play a major role in innate immune defense in the lung.

SP-B is a small protein (a homodimer of 8.7 kD monomers) with approximately 40 % of its amino acids of hydrophobic nature and with 30–45 % α-helical secondary structure. SP-B contains seven **C** residues, six of them forming three intramonomer disulfide bonds and one establishing an intermonomer disulphide bridge that yields a covalent homodimer. The distribution of polar and non-polar amino acids in the α-helices renders them amphipathic along their helical axis.

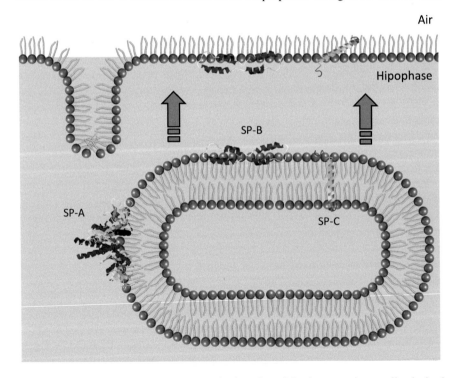

Fig. 2.8 Current models on the structure and orientation of the three proteins usually obtained associated with pulmonary surfactant membranes, SP-A, SP-B and SP-C. Adapted by permission of from Pérez-Gill (2008)

This sort of arrangement is often found in proteins that interact peripherally with membranes or which can form hydrophilic channels in membranes. SP-B appears to behave like a tightly bound peripheral membrane protein. It forms two types of interactions with phospholipids. The basic residues of the SP-B helices are proposed to form ionic interactions with PG in the phospholipids, and the hydrophobic portions of the helices interact with the hydrophobic parts of the phospholipids. These interactions are thought to be essential for the role of SP-B in the surfactant interfacial adsorption of phospholipids.

SP-C is also a small hydrophobic protein (3.7 kDa) with a α-helical secondary structure (about 70 %) as well (see Fig. 2.8). The N-terminal segment of SP-C has a positive net charge and includes two palmitoylated cysteine residues. These palmitic acyl chains are essential to anchor the N-terminal segment of SP-C into the membrane. The C–terminal region is enriched in branched aliphatic residues forming an α-helical motif, which spans the bilayer. In the membrane, SP-C is tilted ~70° respect to the membrane plane, exposing the positive charges and allowing for the establishment of preferential interactions with anionic phospholipids.

SP-A and SP-D are hydrophilic proteins that are components of the innate immune system. They modulate the inflammatory response involved in removing pathogens from the epithelial surfaces. They recognize numerous types of microorganisms, including viruses, bacteria, and fungi. SP-A and SP-D contain specific regions that are able to bind different carbohydrate domains present on pathogenics surfaces. In addition, SP-A can bind DPPC through these regions. This is an interaction that is suggested to be critical for the formation of tubular myelin (a unique and regular network of membranes extended by the surfactant at the airways). The N-terminal sequences of SP-A and SP-D seem to be necessary not only for oligomer stabilization, but also for the interaction with phospholipids and the formation of tubular myelin.

2.5 Structures of Lipids in Aqueous Environments

Phospholipids are said to be mesomorphic in that they can exist in several organizational forms, both in pure forms and when they are dispersed in aqueous environments. As discussed in Chap. 1, all phospholipids are amphiphilic compounds that contain a polar region and a hydrophobic region, but their behavior in aqueous environments varies considerably (Gennis 1989). They take up different phases wherein the fundamental lipid structures can be simplified into shapes as shown in Fig. 2.9.

While lipids exhibiting an inverted truncated cone-shape with small head group tend to adopt the hexagonal inverted structure, H_{II}, species such as lysophospholipids and many detergents, with a relative small headgroup and often with only one acyl chain, tend to adopt a cone shape, H_I. In turn, lipid molecules with cylindrical shape, such as PC or SM, are likely to adopt the structure of a bilayer. Intralipid moieties such as the nature of the polar group and the chain length and degree of acyl chain unsaturation will influence the shape and the tendency to

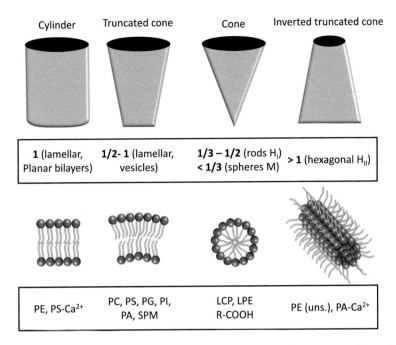

Fig. 2.9 Molecular shapes and the critical packing parameter for some membrane lipids. Redrawn after Gennis (1989)

form different phases. A so-called critical packing parameter (P) was introduced to rationalize this phenomenological behavior (Israelachvili et al. 1976)

$$P = \frac{v}{S_o l_c} \qquad (2.10)$$

where v is the molecular volume, S_o the optimal or equilibrium surface area of the polar group at the interface of the aggregate and l_c the chain length. P can then be used to predict the preferred association structure based on the geometrical parameters of the amphiphile. When $P \leq 1/3$, a micellar sphere is expected; when $1/3 \leq P \leq 1/2$ cylindrical micelles are likely to accur; when $1/2 \leq P \leq 1$ the formation of bilayers is favored and when $P > 1$, H_{II} phases are formed.

P is a geometrical parameter expressing the intrinsic curvature of a phospholipid species as defined in 2.10 by the ratio of the diameter of the head group to that of the acyl chains. As we will discuss in Sects. 3.4 and 3.5 the spontaneous curvature of lipids, referring to their natural tendency to bend in the most thermodynamically stable conformation, plays a crucial role in the lipid-IMP interaction. The spontaneous curvature, indicated by c_o and expressed in nm^{-1} describes the tendency to adopt a curved conformation, positive or negative and is dependent of the headgroup-acyl chain balance. $c_o = 0$ for cylindrical formed lipids, $c_o < 0$ for lipids with tail region bigger than the headgroup and, conversely, $c_o > 0$ for lipids with tail region smaller than the headgroup. For instance, negative c_o values will be found for POPE, DOPE or CHOL and positive c_o values will be found for POPC, DPPC or DSPC. As will be discussed in Chap. 4, it is known that the curvature of lipid monolayers plays a crucial role in the activity of several IMPs .

2.5.1 Hydrophobic Effect and Lipid Self-aggregates

The solubility of lipids in water is very low due to unfavorable hydrophobic interactions of the acyl chains with water. The solubility depends on the temperature, and the length and size of the acyl chains. Many surfactants, such as soaps detergents or bile salts, have a significant solubility in water as monomers. When their concentrations are increased, they spontaneously form small aggregates called micelles, which can exist in equilibrium with the monomers. The formation of micelles occurs for each amphiphile at a defined solute concentration, called the critical micellar concentration (CMC). Below the CMC the amphiphilic molecules tend to migrate to the air-water interface (Fig. 2.10) with the acyl chains oriented to the less apolar phase, that is, the air. For other amphipathic compounds with a higher hydrophobic to hydrophilic nature such as a phospholipid, monomer concentrations can be vanishly small, and larger aggregate structures are formed at very low concentrations.

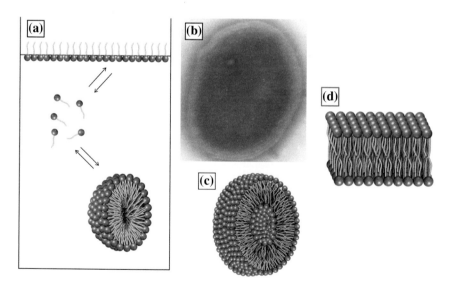

Fig. 2.10 At small concentrations of surfactant, monomers tend to migrate to the air water interface and when the concentration is ~CMC, micelles are formed beneath the monolayer (**a**). Multilamellar liposomes stained with phosphotungstic acid as observed by EM (**b**); cartoon of a unilamellar liposomes (**c**); and a planar lipid bilayer (**d**)

The aggregation into these macromolecular structures results from the hydro-phobic effect. This refers to the association of the hydrophobic moieties of mol-ecules in order to avoid an aqueous environment. This leads to the formation of structures with the hydrophilic headgroups facing the aqueous environment and the hydrophobic groups in self-aggregated arrays.

Segregation of amphiphiles into micelles or bilayers can be described by means of thermodynamic analysis. The chemical potential of an amphiphile can be expressed as

$$\mu_i = \mu_i^\circ + kTLn\chi_i \tag{2.11}$$

where μ_i° is the standard state chemical potential of species i in solution, χ_i the mole fraction, k the Boltzmann constant and T the temperature. In a solution of amphiphiles we can find several aggregation states in equilibrium, from the mono-meric form to the structure with the maximum number (N) of aggregated mole-cules. Hence, the chemical potential can be expressed as

$$\mu_N = \mu_N^\circ + \frac{\kappa T}{N}Ln\left(\frac{\chi_N}{N}\right) \tag{2.12}$$

where μ_N° is the standard state chemical potential of the species with a number N of molecules. Three terms have been postulated (Gennis 1989) to contribute to the value of μ_N° according the following equation

$$\mu_N^\circ = \gamma S + \frac{C}{S} + G_N \tag{2.13}$$

where γ is the interfacial surface tension, S the average surface area, G_N the Gibbs energy associated with the acyl chains and C is a constant. The first two terms are attractive and repulsive, respectively, whilst the third one is defined as an energetic term associated with the acyl chains. The first two terms account for the ener-getic contribution due to the intermolecular interactions at the water-hydrocarbon region. Notice that the optimal surface area (S_o) can be easily obtained by setting $d\mu_N^\circ/dS = 0$ which gives $S_o = (C/\gamma)^{1/2}$. This relationship shows that the value of S_o depend on C, the repulsion constant, in such a way that larger values of C result in larger values of S_o.

If we assume an equilibrium between the monomeric state ($N = 1$) and the micelle state with the maximum number of aggregation N, we can accordingly write from the phase equilibrium rule that

$$\mu_1^\circ + kTLn\chi_1 = \mu_N^\circ + \frac{kT}{N}Ln\left(\frac{\chi_N}{N}\right) \tag{2.14}$$

and rearranged it in the following way

$$\mu_N^\circ - \mu_1^\circ = kTLn\chi_1 - \frac{kT}{N}Ln\frac{\chi_N}{N} \tag{2.15}$$

The difference between the standard states of the chemical potentials can be inter-preted as the hydrophobic Gibbs energy because of the exclusion of water from the apolar region of the amphiphile when the aggregate is formed. More intuitively, $\mu_N^\circ - \mu_1^\circ$ is the Gibbs energy required to transfer a monomer to the aggregate in such a way that the more negative this energy is the smaller χ_N will be. In other words, from Eq. 2.15 it can be inferred that the concentration of the lipid in the aggregated form, the χ_N increases when the difference in the standard chemical potentials increase.

2.5.2 Liposomes

Liposomes were initially described as closed structures consisting on a large number of concentric bilayers that were formed upon the hydration of phospholipids (Bangham et al. 1965). Over the subsequent 50 years the great utility of liposomes prepared by the hydration of natural or synthetic phospholipids as model membranes and drug delivery systems, among other uses has been demonstrated by their use in tens, if not hundreds, of thousands of experiments. When dry phospholipids are dispersed in water, multilamellar vesicles (MLVs) (Fig. 2.10) are formed spontaneously. By using the appropriate techniques the size of the liposomes can be controlled. Typically they are classified as small unilamellar vesicles (SUVs; 20–100 nm), large unilamellar vesicles (LUVs; 100–500 nm) and giant unilamellar vesicles (GUVs; 0.5–100 μm).

2.5.3 Supported Membrane Systems

The increasing interest in confining lipid bilayers on solid surfaces for investigation has led to a number of supported membrane systems, including supported lipid bilayers (SLBs), polymer-cushioned lipid bilayers, hybrid bilayers, tethered bilayers, suspended lipid bilayers and supported vesicular layers (Groves and Boxer 2002; Tanaka and Sackmann 2005). In addition to the above mentioned Langmuir-Blodgett (LB) and Langmuir-Schaefer (LS) double deposition methods (Petty 1996), the foremost methods for obtaining SLBs involve the spreading of liposomes and the spin coating of lipids onto preconditioned supports (Simonsen and Bagatolli 2004) (Table 2.1).

2.5.4 Giant Unilamellar Vesicles (GUVs)

Giant unilamellar vesicles (GUVs) (Fig. 2.11) have been widerly used in the study of membrane nanomechanics, and more recently in the investigation of lateral segregation of membrane components. GUVs have diameters between 5 and 200 micrometers and for this reason are suitable for optical techniques such as

Table 2.1 Summary of the common methods used to prepare SLBs: Langmuir-Blodgett/ Langmuir Schaeffer deposition, vesicle fusion and spin coating, together with the main advantages and disadvantages of each technique. The table provides selected references for further information and a schematic diagram of the corresponding method (Picas et al. 2012)

General methods to prepare SLBs			
	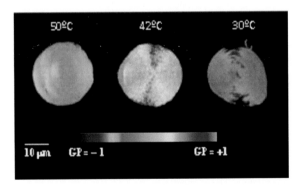		
	Langmuir-Blodgett/ Langmuir Schaeffer deposition	Vesicle fusion	Spin coating
Advantages	Asymmetric bilayers (in composition and lateral pressure)	Simplicity (in terms of manipulation)	Full coverage Absence of defects Single or multi-bilayers
Drawbacks	Leaflet decoupling Uncompleted coverage (TR < 1) Defects (holes)	Symmetric bilayers Equilibrium lateral pressure non controlled	Symmetric bilayers Requires organic solvents Superimposed bilayers often obtained
Refs.	Rinia et al. (1999)	Richter et al. (2003)	Simonsen and Bagatolli (2004)

Fig. 2.11 Two-photon microscopy with Laurdan-labeled DPPCGUVs:GP images of DPPC giant unilamellar vesicles (GUVs) stained with 5 µM Laurdan, in PBS at 50 °C (fluid phase), 42 °C (near the gel to liquid phase transition) and 30 °C (gel phase). Scale bar, 10 µm. GP images were pseudo colored with an arbitrary color palette (Generous gift from Dr. Carlos de la Haba, Dr. José Ramón Palacio, Dr. Paz Martínez and Dr. Antoni Morros)

confocal microscopy. There are several methods of preparation, which yield different sized, shapes and thickness (Bagatolli et al. 2000).

2.5.5 Bilayer Compressibility and Bilayer Surface Pressure

In this section several physical properties derived from the study of the elasticity of the bilayers are introduced. The modulus of compressibility (K_B) of a bilayer is given by

$$K_B = V \left(\frac{\partial P}{\partial V} \right)_{T,n}$$

(2.16)

This expresses the response of the bilayer volume (V) to the pressure (P) at constant temperature (T) and composition (n). Similarly the modulus of surface compressibility (K_A) is given by

$$K_A = A \left(\frac{\partial \overline{T}}{\partial A} \right)_{T,n}$$

(2.17)

that gives the response of the bilayer surface area (A) to a tension ($\overline{T}$) isotropically applied at constant T and n. Typical values of K_B and K_A for a fluid bilayer are $(1–3) \times 10^9$ and 0.14 N m^{-1} respectively.

The curvature elasticity (B) refers to the change of curvature of a surface in response to a bending moment (M) acting on the bilayer edge

$$B = \frac{\partial M}{\partial (1/R)} \bigg|_T$$

(2.18)

An important magnitude is the surface bilayer pressure (π_B), that, contrarily to the surface monolayer pressure (π), cannot be obtained experimentally. Indeed, it is interesting to know the equivalence between π and π_B. This can be done by investigating the isotherms of a given lipid a several temperatures and determining the T_m of the lipid of interest by plotting the molecular surface area as a function of the temperature and taking the maximum slope at a given pressure (Blume 1979). By using this method one finds that the $\pi = \pi_B = 30$ mN m^{-1}. In this regard, using surface thermodynamics, one can write the equation for the mechanical equilibrium of the bilayer membrane as

$$\overline{T} = \gamma - \pi$$

(2.19)

where γ is the interfacial Gibbs energy density (equivalent to the hydrophobic surface energy density) and π is, the surface pressure in the bilayer. Since for large vesicles the isotropic tension has been demonstrate to be zero, (2.19) becomes

$$\gamma = \pi_0$$

(2.20)

Then, taking a value of 70 mN m^{-1} for the hydrophobic Gibbs energy, one obtains a value for π_0 for each of the two halves of the monolayer of 35 mN m^{-1}, which is in good agreement with the value of π_B discussed above.

2.6 The Lipid-Phase Transition: Some Experimental Approaches

2.6.1 Differential Scanning Calorimetry of Lipids

In the presence of an excess of water, phospholipids form fully solvated lipid bilayers, which undergo a phase transition from the solid-like gel state (L_α) to the fluid liquid-crystalline state (called L_α) or liquid-disordered (L_d) state at a temperature (T_m) that is characteristic for each species and mixture. The value of T_m depends on the acyl chains composition, degree of unsaturation and the headgroup of the phospholipid. DSC measures heat flow into a sample as it is heated at a controlled rate over a range of temperatures, and can be used to measure the phase changes of liposomes. For DMPC and DPPC common phospholipids used in the laboratory, the values of T_m are 24 and 42 °C, respectively. The presence of an unsaturated acyl chain in the phospholipid structure produces a decrease in the T_m. Figure 2.12 shows the endotherms for a series of heteroacid phospholipids with the same headgroup, PC, a palmitic acid esterified at the sn-1 position and different unsaturated acyl chains at the sn-2 position (Hernandez-Borrell and Keough 1993). As can be seen POPC and PSPC show narrow endotherms with their T_m at −2 and 49 °C respectively. By increasing the number of double bonds from 1 to 6, the endotherm becomes more complex and the T_m value is decreased.

The lamellar L_α (liquid crystalline) state is the most common state found in biological membranes and it is characterized by relative disorder in the acyl chain region in comparison to the lamellar L_β (gel) state. Most natural membranes contain a significant amount of unsaturated phospholipids that mixed with the other components maintain the natural membrane in the L_α state. The transition to H_{II} phases can also be detected by using DSC. An example is the endotherm of POPE in presence of Ca^{2+} shown in Fig. 2.18a, where the transitions from L_β to the L_α and from L_α to the H_{II} are observed at 24.6 and 59.4 °C, respectively.

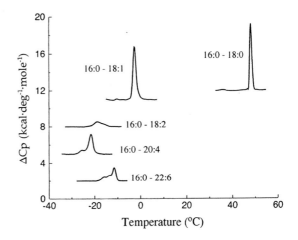

Fig. 2.12 Endotherms for liposomes of PSPC (16:0–18:0), POPC (16:0–18:1), PLPC (16:0–18:2), PAPC (16:0–20:4) and PDPC (16:0–22:6). Redrawn from Hernandez-Borrell and Keough (1993) with permission from Elsevier Science

In the case of eukaryotic cell membranes, the presence of cholesterol in a given proportion results in bilayers showing intermediate properties between those of L_β and L_α state, and for this reason the concept of a liquid-ordered (L_o) phase has been introduced. This terminology is frequently associated with the literature on "Rafts" or detergent resistant membrane domains (DRMs) and when describing lipid phases of the lung surfactant.

Cholesterol causes a broadening of the endotherm and a decrease in the enthalpy of the transition whilst the T_m is gradually moved to lower values. The measureable change in the value of the enthalpy decreases as cholesterol concentration increases. The broadening of the transition and the loss of enthalpy correspond to a decrease in broad range of temperature. The amount of cholesterol required for the abolition of the enthalpy depends on each phospholipid species. In Fig. 2.13, for instance, 30% and 20 % mol/mol of cholesterol is required to abolish the enthalpy of the transition of PSPC and POPC, respectively. The thermal behavior of lipids is critical in the formation of vesicles in solution. Liposomes are normally prepared at a temperature above of the T_m of the selected phospholipid species because below its T_m the hydration of the lipid and its formation into liposomes is extremely (orders of magnitude) slower than for lipids above T_m. For this reason this process of forming closed vesicles may take minutes when the lipids are above T_m and even after hours of hydration below T_m fully sealed vesicles cannot be obtained.

2.6.2   Fluorescence Anisotropy

This method consists of measuring the anisotropic emission of fluorescent labels that are incorporated in the membranes. The fluorescence anisotropy values are calculated according to

$$r = \frac{I_{VV} - GI_{HV}}{I_{VV} + 2GI_{HV}} \tag{2.21}$$

where I_{VV} and I_{HV} are the intensities measured when polarizers are parallel and perpendicular to the exciting beam and G is the grating correction factor equal to I_{HV}/I_{HH}. In this technique, lipid-soluble fluorophores such as DPH (1,6-diphenyl-1,3,5-hexatriene) and TMA-DPH (trimethylammonium-DPH) with the ability to intercalate within the lipid bilayers are often used. DPH is a non-polar, but polarizable polyene that was the first molecule used to describe membrane fluidity/microviscosity. Fluidity is very low when the acyl chains are in the L_β phase and increases when the bilayer reaches the specific T_m. The relationship between the viscosity at a given temperature $\eta(T)$ is given by

$$\eta(T) = \eta(T_m)^{-\Delta E\theta/RT} \tag{2.22}$$

where $\theta = (T - T_m)/T_m$ is the reduced temperature and ΔE the change in the energy during the transition.

While DPH shows a negligible fluorescence in water when it is added to lipid membranes it intercalates spontaneously within the acyl chains resulting in an increase of its fluorescence. When inserted in bilayers, DPH may exist in two orientations: one with its long axis parallel to the lipid acyl chains axes; and the other in the core of the bilayer, with its long axis parallel to the bilayer surface. DPH does not show preference either for the gel or for the fluid (liquid crystalline) phase. For TMA-DPH the DPH moiety is inserted between the lipid acyl chains, but due to its net positive charge, the TMA group lies in the water-lipid interface of the lipid bilayer, near the phospholipid head groups. When the lipid matrix in which the probe is embedded undergoes a transition between two phases, the anisotropy changes dramatically due to an increase (lower anisotropy) or decrease (higher anisotropy) of the membrane fluidity. These changes are easily measured using a conventional steady-state fluorimeter equipped with polarizers. Fluidity changes in different parts of the lipid bilayer due to phase transitions can also be tracked using different labels that show different localization in the membrane (as for example, DPH and TMA-DPH). In some cases, as in the study of the annular region around IMPs (Sect. 3.3) fluidity can also be assessed by using pyrene labeled phospholipids, by calculating the excimer-to-monomer ratio of this molecule. Pyrene is excited at 338 nm and two fluorescence maxima are obtained, at 375 and 470 nm, corresponding to the monomer and excimer bands, respectively.

After excitation by polarized light, the fluorescent dye experiences a free rotation during its fluorescent lifetime that is highly dependent on the membrane viscosity. Therefore, the degree of depolarization of the fluorescent emission versus the excitation light is intimately related with the degree of motion or fluidity in the lipid bilayer or, more precisely, with the fluidity of the microenvironment where the fluorophor is located. The fluorescent anisotropy value (r) is calculated from the intensity of fluorescent emissions measured in different directions by using polarizers. Figure 2.14 shows the phase transition of synthetic lipid POPE:POPG (3:1, mol/mol) by following the fluorescent anisotropy decay of DPH at increasing temperatures. A sharp phase transition from L_β to L_α can be observed corresponding to a sudden change in membrane fluidity. The following equation can be fitted to the anisotropy versus temperature data

$$r = r_1 + p_1 T + \frac{r_{s_1} - r_{s_2} + p_2 T - p_1 T}{1 + 10^{B\left(\frac{1}{T} - \frac{1}{T_m}\right)}} \qquad (2.23)$$

where T is the absolute temperature, B is a measure of the cooperativity of the transition, p_1 and p_2 correspond to the slopes of the straight lines at the beginning and at the end of the plot, and r_1 and r_2 are the anisotropy intercepting values at the y axis. From r values the limiting anisotropy (r_∞) was determined using the following relationship

$$r^\infty = \frac{4}{3} r - 0.10 \qquad (2.24)$$

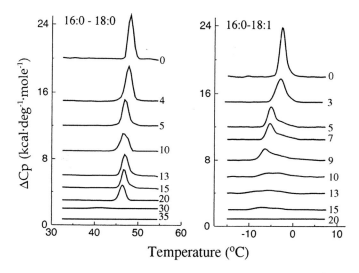

Fig. 2.13 Endotherms of PSPC (*left*) POPC (*right*) liposomes in the presence of different amounts of cholesterol. Redrawn from Hernandez-Borrell and Keough (1993) with permission from Elsevier Science

Fig. 2.14 Typical anisotropy curve of POPE:POPG (3:1) obtained by using DPH as a label

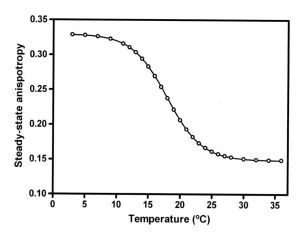

r_∞ reflects restriction of probe motion and can be converted to an order parameter (*S*)

$$S^2 = \frac{r^\infty}{r^0} \tag{2.25}$$

where r^0 is the fluorescence anisotropy in the absence of any rotational motion of the probe.

2.6.3 ^{31}P-Nuclear Magnetic Resonance Spectroscopy

^{31}P-Nuclear magnetic resonance (^{31}P-NMR) spectroscopy provides a suitable technique in the study of the structural organization of biological membrane systems. This approach has the advantage that analysis is not disturbed by the presence of other non-phosphorus probe components. ^{31}P-NMR has been successfully applied to study the conformation and dynamics of phospholipid head groups and also to detect lipid polymorphism.

NMR is sensitive to the angle that the molecules adopt with respect to the magnetic field. As a consequence, anisotropic samples produce broad signals. In order to reduce this effect, samples are submitted to rotation with respect an axis which forms an angle of 54.74° with respect to the magnetic field. This is the so-called "magic angle" (MAS - Magic Angle Spinning) used in "solid-state" NMR. Membranes are anisotropic and moreover their samples possess a certain viscosity, originated by restriction in the molecular motion, which affects relaxation times. This is why most common phospholipids yield broad signals in their ^{31}P-NMR spectra, with spectral widths ranging from 20 to 300 ppm depending on the mesophase and hydration state of the sample.

In Fig. 2.15 the MAS ^{13}P-NMR spectrum of MLVs of the saturated phospholipid DPPC registered at two temperatures is shown, The temperatures were chosen to be below (20 °C) and above (54 °C) the T_m of the phospholipid. A characteristic low field shoulder is visible mainly in the ^{31}P-NMR spectrum for MLVs at the lower temperature. The transition from the L_β to the L_α phase results in a spectral narrowing, which results in a decrease of the chemical shift dispersion, which is often attributed to the increase in the mobility of the phosphate groups. In the fluid state the observed narrower lineshape is attributed to the rapid rotation of lipids around the long axis of the molecule, which have a restricted motion around axes perpendicular to the rotating axis. These motional characteristics are determined by the disposition of the phospholipid molecules in a bilayer.

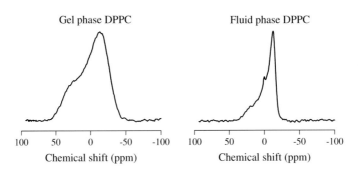

Fig. 2.15 ^{31}P-NMR spectra in the gel-phase (20 °C) and in fluid phase (54 °C) of DPPC multilamellar vesicles (MLV). Spectra were normalized to the same height. Unpublished results generously provided by Dr. Antoni Morros

Therefore, MAS ^{31}P-NMR can be used as a suitable technique to detect the existence of a bilayer structure as well as to monitor the L_β to L_α transition.

The chemical shift anisotropy (CSA) of the phosphorus atoms in phospholipids is a parameter frequently used to characterize the ^{31}P-NMR spectra of phospholipids. It is calculated as the chemical shift difference between the high field peak and the low field shoulder. Since the ^{31}P-NMR spectra of mixed liposomes do not always show a well-defined shoulder at low field, the determination of CSA directly from the powder pattern may not be accurate. Alternatively, the second moment (M_2) of the powder pattern spectra, which can be accurately evaluated (Herreros et al. 2000), is determined. M_2 values report on the square of the headgroup order parameter and so reflect the changes occurring in the structure and dynamics of the phosphate headgroups (Léonard and Dufourc 1991; Gaillard et al. 1991). It is defined as

$$M_2 = \frac{\sum (w_i - M_1)^2}{\sum I_i} \qquad (2.26)$$

where w_i and I_i are the frequency and intensity of the ith data point, respectively, and M_1 is the first moment, which corresponds to the isotropic chemical shift of the sample. The short recycling delay used in these experiments (TR = 1.2 s) results in small differential saturation effects, that in turn translate into moderately overestimated M_2 values with respect to those calculated from fully relaxed spectra (TR > 3 s).

Figure 2.16 shows the temperature dependence of the second spectral moment (M_2) for DPPC. For DPPC a temperature increase leads to a monotonic decrease in M_2 followed by a drastic decrease corresponding to the main transition temperature of the lipid bilayer which can be attributed to an increase in the local motions of phospholipid headgroups.

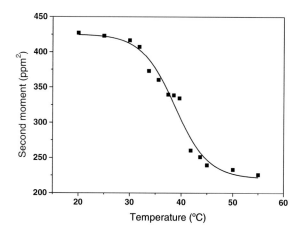

Fig. 2.16 Temperature dependence of the second moment (M_2) calculated from ^{31}P-NMR spectra of DPPC multilamellar vesicles (MLV). Unpublished results generously provided by Dr. Antoni Morros

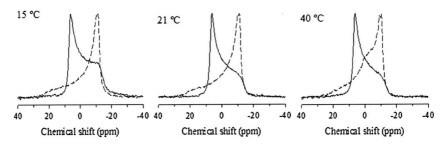

Fig. 2.17 Solid-state ^{31}P-NMR spectra at 15, 21 and 40 °C corresponding to POPE:CL (0.8:0.2, mol/mol) dispersions in Ca^{2+} free 50 mM Tris–HCl, pH 7.40, 150 mM NaCl buffer (*dashed line*), and in 20 mM $CaCl_2$ added buffer (*continuous line*). Reprinted from Domènech et al. (2007b) with permission from Elsevier Science

^{31}P-NMR spectroscopy is also an appropriate technique to unambiguously characterize the lamellar to H_{II} phase transition. Figure 2.17 shows the ^{31}P-NMR powder pattern spectra of POPE:CL (0.8:0.2, mol:mol) systems at 15, 21 and 40 °C in the presence (continuous lines) and in the absence (dashed lines) of 20 mM $CaCl_2$. While, samples in the absence of Ca^{2+} show mainly the usual features of the lamellar phase, in the presence of calcium, the spectra show the typical shape corresponding to the H_{II} phase with the chemical shift dispersion having an "opposite" shape compared to the bilayer shape. At 15 °C the contribution of the lamellar phase was observed. This feature progressively disappears when temperature increases. In the H_{II} phase, there is a rapid diffusion of phospholipid along the cylinder axis which represents an additional mechanism of motional averaging (Domènech et al. 2007b).

2.6.4 AFM in Force Spectroscopy (FS) Mode

AFM has been extensively used for topographic characterization of model membranes, and in particular, for identification and visualization of lipid domains and phase transitions in LBs (Oncins et al. 2007; Picas et al. 2008) and SLBs (Domènech et al. 2007a, b, c; Seeger et al. 2009). However, in cases where domains are not distinguished, phase transitions can be followed by monitoring the changes in the roughness of the surface at different temperatures. Accordingly DSC measurement (Fig. 2.18a) POPE SLBs present the L_{β} to the L_{α} transition at 24.6 °C and the L_{α}-H_{II} transition at 59.4 °C. Topography changes changes are clearly apparent for these SLBs as shown in Fig. 2.18b.

AFM in force mode measurements are based on cantilever deflection as a function of the piezoelectric tube position where the sample is mounted and which is responsible for the upwards and downwards movement after applying a voltage. Noteworthy, the technique is commonly known as "force spectroscopy" (FS),

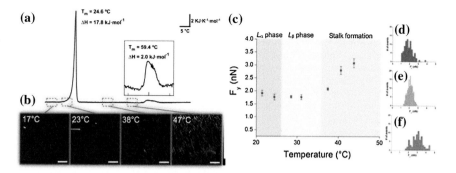

Fig. 2.18 Endotherm of POPE containing 20 mM of $CaCl_2$, displaying the lamellar L_β to L_α transition at 24.6 °C and the lamellar to hexagonal phase (H_{II}) transition at 59.4 °C (**a**). AFM topography images acquired for POPE/POPE (30 mN m^{-1}/25 mN m^{-1}) SLBs at different temperatures (17, 23, 38, and 47 °C) (**b**). Distribution of breakthrough forces (F_y) obtained from FvD curves performed at different temperatures for POPE/POPE (30 mN m^{-1}/25 mN m^{-1}) SLBs (**c**), showing the different mechanical response upon changes in bilayer polymorphism. Representative histograms of the F_y variation as a function of the thermotropic state of POPE/POPE (30 mN m^{-1}/25 mN m^{-1}) SLBs (L_β phase *blue*, L_α phase *grey*, and stalk formation, *green*). Color scale is 20 nm and the bar scale is 1 μm. Adapted from Picas et al. (2008)

even if there is no specific matter-radiation interaction. Two main items of information can be extracted from the FS curves: (i) the breakthrough or threshold force (F_y), i.e. the force that the bilayer can withstand before being indented (ii) the adhesion force (F_{adh}) between the tip and the bilayer. Actually, FS experiments are an interesting approach to reveal the nanomechanics behind the thermotropic behavior of lamellar phases. This introduces the possibility of studying other membrane systems and lipid polymorphisms following a similar approach. Thus the technique allows following the L_β to the L_α and the L_α to the H_{II} transitions of POPE though F_y measurements. In this particular experiment, within the temperature range, evaluation of breakthrough forces indicated a progressive increase in membrane resilience during transition from L_α to H_{II} and the appearance of intermediate structures, known as stalks, at around 45 °C (Fig. 2.18c). F_y values are the average of the distributions shown in Fig. 2.18d–f. By using the same tips and experimental conditions, FS provides a means to unambiguously discriminate betweeen phospholipid species and phases (Garcia-Manyes et al. 2010).

References

Bagatolli LA, Parasassi T, Gratton E. Giant phospholipid vesicles: comparison among the whole lipid sample characteristics using different preparation methods—a two photon fluorescence microscopy study. Chem Phys Lipids. 2000;105:135–47.

Bangham AD, Standish MM, Weissmann G. The action of steroids and streptolysin S on the permeability of phospholipid structures to cations. J Mol Biol. 1965;13:253–9.

Binnig G, Rohrer H, Gerber C, Weibel E. Surface studies by scanning tunneling microscopy. Phys Rev Lett. 1982;49:57–61.

Blume A. A comparative study of the phase transitions of phospholipid bilayers and monolayers. Biochim. Biophys. Acta—Biomembr. 1979;557(1):32–44.

Davies JTRE. Interfacial Phenomena. 1st ed. New York: Academic Press Inc.; 1963.

Domènech Ó, Ignés-Mullol J, Teresa Montero M, Hernandez-Borrell J. Unveiling a complex phase transition in monolayers of a phospholipid from the annular region of transmembrane proteins. J Phys Chem B. 2007a;111(37):10946–51.

Domènech Ò, Morros A, Cabañas ME, Teresa Montero M, Hernández-Borrell J. Supported planar bilayers from hexagonal phases. Biochim Biophys Acta—Biomembr. 2007b;1768(1):100–6.

Domènech Ò, Redondo L, Picas L, Morros A, Montero MT, Hernández-Borrell J. Atomic force microscopy characterization of supported planar bilayers that mimic the mitochondrial inner membrane. J Mol Recognit. 2007c;546–53.

Gaillard S, Renou JP, Bonnet M, Vignon X, Dufourc EJ. Halothane-Induced Membrane Reorganization Monitored by Dsc, Freeze-Fracture Electron-Microscopy and P-31-Nmr Techniques. Eur Biophys J. 1991;265–74.

Garcia-Manyes S, Redondo-Morata L, Oncins G, Sanz F. Nanomechanics of lipid bilayers: Heads or tails? J Am Chem Soc. 2010;132(14):12874–86.

Gennis RB. Biomembranes : molecular structure and function. Barcelona [etc.] : Springer; 1989.

Groves JT, Boxer SG. Micropattern formation in supported lipid membranes. Acc Chem Res. 2002;35:149–57.

Hernandez-Borrell J, Keough KM. Heteroacid phosphatidylcholines with different amounts of unsaturation respond differently to cholesterol. Biochim Biophys Acta. 1993;1153(2):277–82.

Herreros B, Metz AW, Harbison GS. Moment analysis as a systematic tool for NMR powder pattern analysis. Solid State Nucl Magn Reson. 2000;16:141–50.

Hoenig D, Moebius D. Direct visualization of monolayers at the air-water interface by Brewster angle microscopy. J Phys Chem [Internet]. 1991;95:4590–2. Available from: http://pubs.acs.org/doi/abs/10.1021/j100165a003.

Israelachvili JN, Mitchell DJ, Ninham BW. Theory of self-assembly of hydrocarbon amphiphiles into micelles and bilayers. J Chem Soc Faraday Trans. 1976; 2:1525.

Léonard A, Dufourc EJ. Interactions of cholesterol with the membrane lipid matrix. A solid state NMR approach. Biochimie. 1991;73:1295–302.

Maget-Dana R. The monolayer technique: a potent tool for studying the interfacial properties of antimicrobial and membrane-lytic peptides and their interactions with lipid membranes. Biochim Biophys. Acta—Biomembr. 1999.

Nag K, Keough KM. Epifluorescence microscopic studies of monolayers containing mixtures of dioleoyl- and dipalmitoylphosphatidylcholines. Biophys J. 1993;65(3):1019–26.

Oncins G, Oncins G, Picas L, Picas L, Hernandez-Borrell J, Hernandez-Borrell J, et al. Thermal response of Langmuir-Blodgett films of dipalmitoylphosphatidylcholine studied by atomic force microscopy and force spectroscopy. Biophys J. 2007;93:2713–25.

Pérez-Gil J. Structure of pulmonary surfactant membranes and films: The role of proteins and lipid-protein interactions. Biochim. Biophys. Acta—Biomembr. 2008;1778:1676–95.

Peters R, Beck K. Translational diffusion in phospholipid monolayers measured by fluorescence microphotolysis. Proc Natl Acad Sci. 1983;80(23):7183–7.

Petty MCM. Langmuir-Blodgett films: an introduction. Cambridge: Cambridge Univ Press; 1996.

Picas L, Milhiet PE, Hernández-Borrell J. Atomic force microscopy: a versatile tool to probe the physical and chemical properties of supported membranes at the nanoscale. Chem Phys Lipids. 2012;845–60.

Richter RP, Him JLK, Brisson A. Supported lipid membranes. Mater Today. 2003; 32–7.

Rinia HA, Demel RA, van der Eerden JP, de Kruijff B. Blistering of langmuir-blodgett bilayers containing anionic phospholipids as observed by atomic force microscopy. Biophys J. 1999;77:1683–93.

Schurch S, Lee M, Gehr P. Pulmonary surfactant: Surface properties and function of alveolar and airway surfactant. Pure Appl Chem. 1992;1745–50.

Seeger HM, Marino G, Alessandrinia, Facci P. Effect of physical parameters on the main phase transition of supported lipid bilayers. Biophys J Biophys Soc. 2009;97(4):1067–76.

Simonsen AC, Bagatolli LA. Structure of spin-coated lipid films and domain formation in supported membranes formed by hydration. Langmuir. 2004;20:9720–8.

Tanaka M, Sackmann E. Polymer-supported membranes as models of the cell surface. Nature. 2005;437:656–63.

Verger R, Mieras MCE, De Haas GH. Action of phospholipase A at interfaces. J Biol Chem. 1973;248:4023–34.

Chapter 3
Lateral Distribution of Membrane Components and Transient Lipid-Protein Structures

Abstract The individual physicochemical properties of membrane phospholipids, acyl chain composition and headgroup charge, are major determinants of lipid phase separation into domains. In this chapter we will review the general procedure to reconstitute integral membrane proteins (IMPs) into supported lipid bilayers (SLBs). Second, a brief introduction to lipid phase diagrams is presented. Then we will discuss the wide variety of lipid-protein structures reported in the literature. Protein affinity for lipids is discussed on the basis of temporal and spatial residence of lipids at the lipid-protein boundary, leading to definitions of boundary, associated non-boundary-lipid, and the remainder of lipids that are distributed in the bulk of the bilayer. We present some examples to illustrate the hydrophobic matching between lipids and IMPs, and based on physical properties of the lipid, stretching and bending, we introduce the surface flexible model (SFM) of the membrane which is consistent with the continuum theory of matter applied to membranes. This model accounts for the experimentally observed physicochemical behaviours to interpret the insertion of IMPs into specific lipid domains of the membrane.

Keywords Lipid phase diagrams · Boundary lipid · Hydrophobic matching · Curvature stress

3.1 Lateral Distribution in Reconstituted Systems

The lateral packing and physical state of the bilayer strongly affect the lateral distribution of IMPs and their dynamic properties such as lateral motion and conformational changes occurring during protein specific function (e.g. enzyme activity or transport). When IMPs are introduced into artificial membranes at low concentrations they tend to be randomly distributed in the lipid matrix and they undergo relatively slow diffusion compared to the lipids (e.g. two-dimensional diffusion coefficient of 11 $\mu m^2 \, s^{-1}$ for the lipids in a DOPC:DOPG matrix, compared to 4 $\mu m^2 \, s^{-1}$ for LacY in the same system). The mobility in natural membranes depends, among other factors, on the degree of protein crowding, the protein size and specifically on the lipid-to-protein (LPR) ratio. These properties have been extensively demonstrated using experimental techniques, typically FRAP or fluorescence correlation spectroscopy (FCS) on GUVs. These techniques as well as molecular dynamics simulations indicate that both rotational and translational diffusional coefficients decrease upon protein clustering or upon an increase in the size of the IMPs. An understanding of the degree of IMP clustering in natural membranes and the biological relevance of lateral organization can be approached by using membrane models into which the extracted and purified IMPs are reconstituted.

In the laboratory, IMPs are extracted from natural membranes with surfactants and afterwards reconstituted into models, where the LPR can be adjusted conveniently for the purposes of each experiment. An IMP surfactant extract is mixed with a suspension of micelles formed with the desired lipids in the same surfactant (Fig. 3.1). The surfactant is removed by dialysis, gel filtration, or, more often, by addition of hydrophobic substrates such as polyestyrene beads to obtain proteoliposomes at the desired LPR. Proteoliposomes or GUVs are convenient models to investigate the lateral organization of the membrane through confocal microscopy. In addition, the lateral organization of unlabeled IMPs in lipid media in proteoliposomes can also be deposited onto a flat surface to obtain proteolipid sheets (PLSs) that can be then observed by AFM or other applicable surface techniques.

An alternative method has been proposed which consists in adding the extracted IMP onto previously destabilized SLBs (Milhiet et al. 2006). In this regard the selection of the adequate surfactant is a delicate matter. As can be seen in Fig. 3.2, the effects of N-octyl-β-D-glucoside (OG), dodecyl-β-D-maltoside

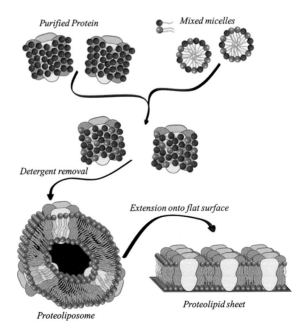

Fig. 3.1 Basic steps in the reconstitution of integral membrane proteins into liposomes and supported lipid bilayers: Formation of proteolipid sheets

Purified Protein

Mixed micelles

Detergent removal

Extension onto flat surface

Proteoliposome

Proteolipid sheet

Fig. 3.2 Interactions of SLB with sugar-based detergents at concentrations above the cmc. **a** Contact-mode imaging of DOPC/DPPC SLB. Lipid phase separation was observed between the darker DOPC fluid phase and the DPPC gel phase. **b** SLB after incubation at room temperature with 0.075 mM (1.5 × cmc) DOTM. Both fluid (*1*) and gel (*2*) phases were conserved; the *darker areas* were the mica, and the *white dots* nonfused vesicles. **c** and **d** SLB after incubation at room temperature (**c**) with 0.3 mM (1.7 × cmc) DDM and (**d**) with 17 mM (1 × cmc) OG. **e** SLB after incubation at 4 °C with 0.075 mM DOTM. The *z* color scale is 15 nm, and *scale bars* are 1 μm. Reprinted from Milhiet et al. (2006) with permission from Elsevier Science

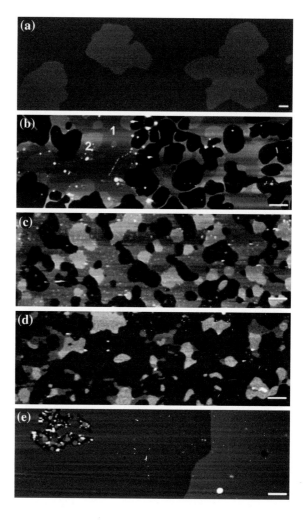

(DDM) or dodecyl-β-s-thiomaltosides (DOTM) on supported lipid bilayers of DPPC:DOPC (1:1, mol/mol) are different. The basic idea behind this method is to empirically find the most resistant SLB while also indicing some surface defects to enhance the protein insertion. In the experiments reported by Milhiet's group, DM and DOTM have been shown to be the more suitable for IMP reconstitution.

3.2 Lipid Phase Separation and Phase Diagrams of Lipid Mixtures

The existence of laterally segregated domains at mico- and nano-scales is inherent to the currently accepted F-MMM (Fig. 1.18). Intriguing questions are whether the lipid domains exist independent of the protein presence, based on non-ideal mixing properties that depend on the diverse structures of the phospholipids only (see Sect. 2.1.3); or, vice versa, the IMPs are inducing the organization of the phospholipids into domains (Poveda et al. 2008).

As discussed in Sect. 2.6, mixed lipid membranes experience relatively broad transitions between the two lamellar phases, L_α and L_β, and also between lamellar and H_{II}, H_I phases. Those transitions reflect the existence of lateral phase separations and, most importantly, the co-existence of micro- and nano-domains of lipids that in turn are dependent on lipid composition, temperature and pressure. Due to the enormous number of possible chemical variations that may occur (acyl chains and/or headgroup differences), phase transitions can be observed in bilayers consisting of mixtures of phospholipids that are heated or cooled over an appropriate temperature range. In this regard binary and ternary phase diagrams are useful in investigating and interpreting lipid phase separations. Classical thermal techniques applied to the study of bilayers such as DSC reach practical limits in interpretation when the number of components in the bilayer is above three. However, spectroscopic and fluorescence techniques allow for additional understanding of phase separations in membranes. It is noted that the presence of other components, such as cholesterol in bilayers induces additional phases or "intermediate" order states that are referred to L_o and L_d states.

Phase diagrams are graphical temperature-composition or pressure-composition representations from which the physical states of substances can be predicted under specific conditions. The Gibbs phase rule states that the number of intrinsic themodynamic variables (F, the "degrees of freedom") that can vary independently at equilibrium is $F = C + 2 - P$, where C and P account for the number of components and phases respectively. In two component systems $F = 4 - P$, and F can be 1, 2 or 3, depending on the number of phases present. For condensed matter systems, it is customary to write the phase rule as $F' = C + 2 - P - R$, where R is the number of degrees of freedom that for practical reasons is considered constant. When applying the phase rule to bilayer membrane systems, the pressure is assumed to be constant, water which is in great excess is ignored as a component, and R is set equal to 1, thus the phase rule for a two component system becomes $F' = 3 - P$, so the maximum number of coexisting phases is two. Actually, a large number of lipid phase diagrams can be found in the literature (Marsh and Phil 1999; Koynova and Caffrey 2002). The description of the phase diagrams for lipid mixtures may become extremely complex (Fig. 3.3), therefore we will describe here only the basic trends of the most common phase diagrams that will be encountered in the lipid literature.

Figure 3.3a describes the theoretical phase diagram for a two component system with complete miscibility in the liquid-crystalline and gel phases, above point **b** and below point **d**, respectively. Between **b** and **d** there is co-existence of both phases and the proportion of each component can be calculated by applying the lever phase rule. Phase diagrams have been constructed by using a variety of techniques, DSC being the most traditional. Using this technique to construct a phase diagram of a binary mixture the onset (T_{onset}) and completion (T_{offset}) temperatures observed in the endotherms (see for instance Figs. 2.15 and 2.16) are ploted

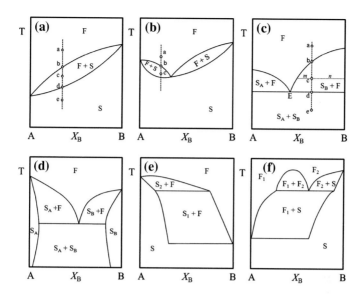

Fig. 3.3 Representative pseudo-phase diagrams for hypothetical phospholipid binary mixtures showing fluid (F) and solid (S) regions

against a series of compositions. An example with similar behavior to the one shown in Fig. 3.3a, is the binary mixture of DPPC with DMPC where the phase diagram has been derived using a number of observational techniques including DSC and the partitioning of the electron spin label TEMPO in the binary system. This fluorescent label has the ability to dissolve in the L_α phase of the membranes but it is excluded from the L_β phase, and this property enables it to be used to follow transitions as a function of temperature. Since DMPC and DPPC share the same headgroup and differ only by two carbons in the length of their acyl chains, deviations from ideal mixing would be expected to be very small, and this is what is observed experimentally. Increasing the difference in acyl chain lengths in the lipid components leads to increasing deviations from ideal mixing in the two fundamental phases and the temperature range of the coexistence region (where the mixtures of gel and liquid-crystalline micro(nano) domains become broader). This broadening of the transition (coexistence) range can be seen in the phase diagrams of the system of DMPC:DPSC (acyl chains difference of four carbons) and for the system of DLPC:DSPC (acyl chains difference of six carbons). Other examples are the mixtures of DSPC:SOPC and DSPC:OSPC for which wide pseudobinary phase diagrams have been reported, or as seen in the much more complex diagram of POPE:POPG, which results from mixtures of a zwitterionic and a negatively charged phospholipid. As the activity coefficients for the mixtures becomes larger, the deviations from ideality lead to diagrams like the one shown in Fig. 3.3b, which features an azeotropic composition of components. Such a behavior has been found for mixtures of 1,2-dielaidoyl-*sn*-glycer-3 phosphocholine (DEPC) (18:1 (Δ9)/18:1 (Δ9)) and DMPC.

Two component mixtures, however, may become more complex when there is ideal miscibility in fluid phase but the gel phase is constituted by the two unmixed pure components. Phase diagrams for such mixtures (Fig. 3.3c) are characterized by the occurrence of a particular composition, the eutectic point, which shows the lowest melting point of any of the mixtures of the two pure components. If we cool down a solution richer in the component B (point **a**) crystals of this component will appear by crossing the boundary region (point **b**). Further cooling (point **c**) and thereafter within the coexistence region results in more extense crystallization of B and a concomitant enrichment of A in the solution (fluid phase). When the eutectic temperature is reached all the solution crystallizes out yielding a crystalline mixture of A and B. Depending on the interactions between the components, phase diagrams may become even more complex. An example of this behavior is seen in mixtures of 1-stearoyl-2-capril-sn-phosphatidylcholine (C:18/C:10) and DMPC (Fig. 3.3d). Indeed the mixtures of these two components result in a phase diagram which is a hallmark of an eutectic system and 3 gel regions below the eutectic temperature (in this case 13.3 °C) (Lin and Huang 1988).

Binary lipid mixtures may have even more complex interactions that give rise to complicated phase diagrams such as the ones shown in Fig. 3.3d–f. The Fig. 3.3e describes the mixing behavior of DOPC:DPPE binary mixtures while Fig. 3.3f exhibits features of the DEPC:DPPE binary mixture as deduced from observations of the temperature dependence of the spin label TEMPO (Hong-wei and McConnell 1975). For DEPC:DPPC mixtures display a phase diagram with regions of complete fluid miscibility or gel miscibility below 70 % mol of DPPE and also immiscibility above this composition. It also displays fluid regions where solid-like mixtures of DEPC and DPPE appear.

Three component lipid mixtures can be investigated through the use of a Gibbs phase triangle. When one of the components is cholesterol this approach can aid in understanding the properties of putative domains termed "lipid rafts" (see Sect. 3.6) that have featured prominently in membrane biophysics and cell biology. For systems with three components, it is common to use a phase diagram consisting of an equilateral triangle where the sides reflect the compositions of the three components. Then the sum of the distances from any point within the triangle drawn perpendicular to the three sides is always equal to 100 when compositions are in mole percents or 1.0 when compositions are in mole fractions. An example of this kind of diagram constructed by using data from four experimental different techniques is shown in Fig. 3.4.

Ternary phase diagrams have been reported for mixtures such as DOPC: PSM:CHOL, DOPC:DPPC:CHOL, POPC:PSM:CHOL and POPC:DPPC:CHOL by determining miscibility transitions using GUVs (Veatch and Keller 2005). In these experiments, coexisting micron-sized domains are observed using fluorescence microscopy over a wide range of compositions and temperatures.

Although they are derived from models of simple composition, binary and ternary phase diagrams are useful in understanding the nature of the lipid bilayer in which IMPs are embedded. An illustration has been reported for the properties of pulmonary surfactant in GUVs (Sect. 2.4.6) as observed by one photon confocal

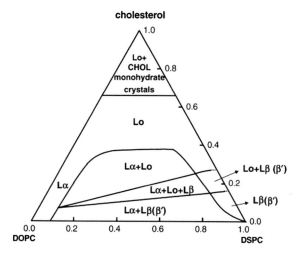

Fig. 3.4 Pseudo-ternary phase diagram for DSPC:DOPC:CHOL Reprinted from Zhao et al. (2007) Biochim. Biophys. Acta J 1768(11):3764–3776 with permission from Elsevier Science

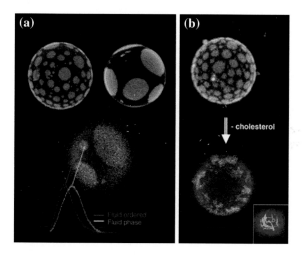

Fig. 3.5 One photon confocal images of GUVs composed of native pulmonary surfactant membranes (*top left*) and DOPC:DPPC:CHOL (*top right*) obtained using the fluorescent probes DiIC18 (*red areas*) and Bodipy-PC (*yellow area*). LAURDAN intensity image of a GUV formed with native pulmonary surfactant: *blue* (L_α) and and *red* (L_o) side contribution of the emission spectrum (A). The extraction of CHOL of GUVs formed by DOPC:DPPC:CHOL by addition of cyclodextrin generates the pattern that compares well with the L_α/L_β coexistence. Reprinted from Bagatolli (2006) Biochim. Biophys. Acta 1758:1541–1556 with permission from Elsevier Science

images for which the ordered and disordered fluid phase coexistence is similar to the one shown by the ternary lipid mixture DOPC:DPPC:CHOL (Fig. 3.5a) (Fidorra et al. 2006).

Intriguingly, while the removal of CHOL from pulmonary surfactant affects the lateral segregation pattern, the extraction of the intrinsic proteins does not (Fig. 3.5b) (De La Serna et al. 2004).

3.3 Lateral Segregation of IMPs: Experimental Evidence

Although there is evidence for the existence of fluid and ordered phase co-existence in some model membranes as well as in some biological membranes, many appear to maintain a generalized near constant "state" of overall order/disorder and intramolecular/intermolecular motions that are termed a global "fluidity" which allows for and influences lateral mobility and conformational changes of the embedded proteins. For example, α-helices of IMPs will not pack easily in phases such as L_β or L_o, and with some exceptions, there is a general preference of IMPs for lipids in the more mobile L_α phase. How structural and other physicochemical factors may modify the equilibrium between the fluid and ordered phases and how the resulting physical state of the bilayer can modulate the activity of IMPs and the lateral distribution of proteins in the plane of the membrane have been the subject of much study. Reconstitution of IMPs in model membranes aims to obtain information on membrane structure and function, and to provide important clues for understanding specific preferences or selectivity between IMPs and lipids.

When IMPs are reconstituted in proteoliposomes, synthetic lipids with low T_m (conveniently below room temperature) are selected to ensure a fluid state for the bilayer. When using a single phospholipid as a matrix for reconstitution, IMPs would distribute randomly within the bilayer if the temperature of the process is kept above the transition temperature of the phospholipids used ($T > T_m$). If the temperature is decreased below the T_m, L_β domains will appear and then IMPs will become mostly laterally segregated into the fluid phase. These behaviors can be demonstrated using, among others, fluorescence spectrophotometry and microscopy and attenuated total reflectance in combination with Fourier transform infrared spectroscopy (ATR-FTIR) or AFM. Some of the early evidence of this behavior arose from the application of freeze-fracture electron microscopy (FF-EM) techniques that allow the direct visualization of IMPs in phospholipid matrices. If the matrix is formed by a binary mixture which does not mix ideally (of particular interest are the ones containing negatively charged phospholipids) the IMPs will also generally show higher affinity for the L_α than for the L_β phase. Notice than when negative phospholipids are present, the presence of divalent cations seems to promote the formation of domains, and clusters, by rigidification of the acidic phospholipids which results in the lateral segregation of lipid domains and clustering of proteins. A classical example is the case of Ca^{2+}-ATPase reconstituted in DOPC and DOPA, where the addition of relatively high concentrations of Mg^{2+} (20 mM) led to the exclusion of the protein from gel domains formed by the complexation of DOPA with the divalent cations, which was demonstrated using FF-EM analysis. In keeping with the idea that the ATPase is separating into the fluid phase in the mixtures of DOPC and DOPA plus calcium, is the finding that in Ca^{2+}-ATPase reconstituted in DMPC, the protein is randomly distributed when the samples are equilibrated above T_m and it is agregated in patches below T_m.

Since natural membranes are usually in a fluid state, in the absence of divalent cations or other physicochemical factors affecting lipid phase separation, membrane proteins will distribute randomly within a bilayer that is formed with natural lipid extracts, or synthethic lipids which are also in the fluid state, i.e., they are at a temperature above their respective T_m. There is extensive literature dealing with this phenomenon, and an early example of that behavior comes from freeze-fracture EM images. In the case of LacY reconstituted in proteoliposomes formed with puridifed lipid extracts from the *E. coli* inner membrane, entities with a diameter of 7.0 ± 1.0 nm (n = 214) were observed randomly distributed inside the convex and concave surfaces of the proteoliposome vesicles (Costello et al. 1984). This size matches reasonably well with the dimensions obtained by X-ray diffraction analysis of 2D and 3D crystals of LacY (Abramson et al. 2003). In this regard AFM is a powerful technique not only because of its high resolution but for giving the possibility of observing biological samples in liquid media. A good example comes from reconstitution experiments of the light-harvesting complex1 (LH1) reaction center (RC) from *Rhodobater sphaeroides* performed in previously formed SLBs destabilized with DOTM (Fig. 3.6). The surfactant induces only a partial solubilization of the DOPC:DPPC (1:1, mol/mol) SLBs in which the LH1-RC has been reconstituted (Fig. 3.6a). When this low magnification image is magnified (Fig. 3.6b) two different topographies are observed: the first is smooth, free of proteins and with a step height difference in respect to the substrate which matches with the thickness of DPPC (~6 nm); and the second is rough and contains bright spots that protrude ~3 nm, which is in fact the estimated height of the cytoplasmic side of the H subunit of the RC. At higher magnification (Fig. 3.6c) we can observe monomeric rings with a mean diameter value of ~9.9 nm, which coincide with estimated value of the RC core complex.

These experiments provide two important additional observations besides the structural information: (i) the IMP is self-segregated into the fluid domains and it is not observed in the gel domain, and (ii) there is a dark and unresolved circle around the RC complex (inset in Fig. 3.6c). The phase diagram of DOPC:DPPC (Fig. 3.6d) (Schmidt et al. 2009) now becomes a tool that we can use to estimate the proportion of each phase at the temperature of the AFM observation. The LH1-RC complex inserts preferentially into the fluid domain, which consists of ~60 % DOPC and ~40 % DPPC at 25 °C. Nevertheless, this is just an approximation because the phase diagram was obtained from liposomes and not from SLBs, which, as we will discuss later affect the T_m of the mixture and hence, the composition of each phase. However, the insertion of the protein will induce a change in the equilibrium distribution of the lipids and the phase diagram will be only approximative.

These observations leave open questions on the level of perturbation in the properties of the lipid bilayers because of the insertion of the IMPs. Furthermore, questions arise about which kind of interactions are mutually affecting lipids and proteins. Several powerful techniques such as X-ray crystallography,

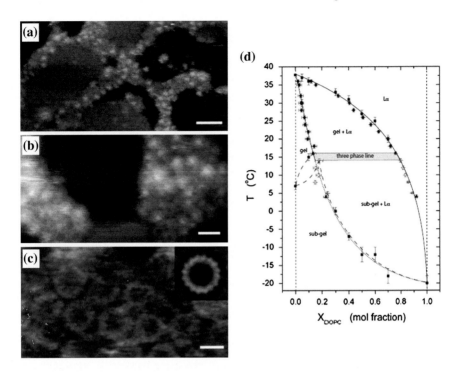

Fig. 3.6 AFM imaging of LH1-RC after direct incorporation into SLB of DOPC:DPPC after 24 h of incubation at 4 °C as observed at room temperature (**a**); higher-magnification (**b**); high-resolution AFM analysis and the 16α/β-heterodimers of LH1 in the average image (inset) (Milhiet et al. 2006); (**c**). Reprinted from Milhiet et al. (2006) with permission from Elsevier Science (**d**). Binary phase diagram of DOPC:DPPC. Reprinted from Schmidt et al. (2009) with permission from AIP Publishing LCC

NMR-spectroscopy, AFM or ATF-FTIR may be applied to monitor these changes, with one of the more traditional ones being DSC. In fact DSC has been used to estimate the minimal size of the segregated domains in nicotin acetylcholine receptor in presence and absence of several lipid system (Poveda et al. 2002). The reduction of the enthalpy as well as the cooperativity of the phase transition with increasing amounts of protein can been observed from DSC endotherms in systems containing glycophorin, lipophilin and the linear peptide gramicidin, among others. Traditionally the decrease in the enthalpy and the loss in the cooperativity of the phase transition is interpreted as a result of the existence of two phospholipid populations, one which is formed by the phospholipids in close contact with the surface of the protein (boundary or boundary lipids); and the other, formed by the unperturbed or bulk phospholipids (see Sect. 3.2) (Houslay and Stanley 1982). We can rationalize the observed behavior in terms of persistence length, the range over which interlipid interactions are somehow affected by new molecular arrangements. This length should become larger when the maximum heat capacity is reached. One would expect that transition cooperativity would decrease as the system becomes more complex—for example, as protein is mixed into the lipid system. One possible interpretation for that behavior is that selected components of the lipid mixture have been "removed" from the transition by their interaction with the protein, and the remaining "free" lipid undergoes a more cooperative transition.

The most powerful evidence of lipid-membrane protein interactions is provided by X-ray crystallography. Although there can be difficulties in the interpretation of electron density maps from proteolipid mixtures that arise because the protein induces unusual conformations of the lipid molecules, there are increasing cases where X-ray information has been obtained from co-crystallized lipids and IMPs. There are many examples of specific lipids, as CL, found in the intramembrane surface of the bacterial photosynthetic center of the AM260 W mutant of *Rb. sphaeroides* and also in KcsA of *Streptomyces lividans* or PE and PG that have been modelled in the Cytochrome c oxidase. Since X-ray crystallography of bound lipids is inherent to the difficulties for extracting the proteins from the natural membranes, we can follow the converse strategy, that is to increase the protein concentration in reconstituted systems to force or enhance the interaction between the IMPs and the phospholipids in the membrane (see Sect. 4.3). While tedious in execution, crystallization in two-dimensions (2D) of IMPs with different lipids might provide clues on specific lipid requirements for appropriate protein insertion within the membrane. While in DSC experiments LPR is ~70, in 2D X-ray experiments LPR can be as low as ~0.5. A well known naturally-occurring 2D crystalline array is that of Br in its membrane environment. In this particular case, minimally invasive treatment of the membrane of *Hallobacterium* (composed of ~75 % Br and ~25 % lipid), yields defined 2D hexagonal lattices. In the case of *H. salinarum,* modelling studies were consistent with several full or partial lipids occupying a large part of the contact surface between monomers in the trimeric organization of the protein. In these arrangements, Br can be visualized by TEM or AFM, as being densely packed into trigonal lattices.

Another means of examining lipid-protein interactions is available through the use of AFM. In series of experiments on LacY, various lipid environments have been used to demonstate how useful this technique can be. Figure 3.7 illustrates different topographic AFM images obtained after the deposition onto mica of proteoliposomes of purified LacY reconstituted in a phospholipid matrix of POPC (Merino et al. 2005). As expected, as the LPR decreases, the tendency for the protein molecules to self-seggregate into protein-enriched domains increases. As can be seen, pure POPC is in a fluid state because of its relatively low transition temperature and low stiffness and it is affected, actually swept away, by the action of the AFM tip (Fig. 3.7a). Conversely, when LacY is present, the mechanical stability of the system increases and the proteolipid layers remain more firmly attached onto the mica substrate. Protrusions, observed as white spots in the images, increase in number as LPR decreases (Fig. 3.7b and c), leading to formation of extensive self-segregated protein patches at an LPR of 0.5 (Fig. 3.7d). A magnified view of these regions reveals the existence of a quasi-crystalline arrangement (Fig. 3.7e) formed by small round entities with a size that corresponds to the

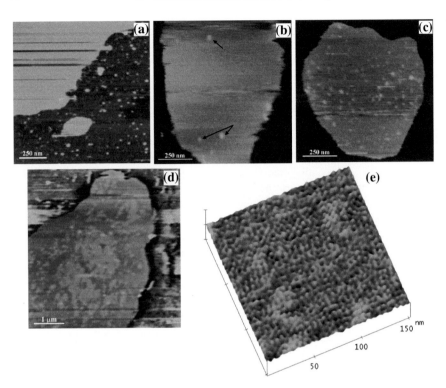

Fig. 3.7 Images obtained using POPC as a lipid matrix and LacY at different values of LPR (w/w): 0 (**a**), 1.5 (**b**), 1.0 (**c**) and 0.5 (**d**). Images were obtained in tapping mode in 10 mM Tris–HCl, 150 mM NaCl (pH 7.40). Protrusions in (b) are indicated by *black arrows*. A high magnification on the aggregated regions of protein of image 3D is shown in (**e**). Round entities with the diameter of LacY can be seen in a 2D quasi-crystalline arrangement. Reprinted from Merino et al. (2005) with permission from Elsevier Science

diameter of the protein obtained from X-ray diffraction. Reducing the LPR facilitates protein self-aggregation and eventually leads to the formation of 2D crystals. It is important to recall that IMPs are normally functional in a monomeric state, and that at least in the case of LacY it has been demonstrated that the activity decreases upon decreasing the LPR ratio.

It is important to note that 2D crystals or segregated domains with IMPs are constituted of a continuous lipid bilayer firmly attached to the protein. This fact points to the existence of natural interactions between lipids and IMPs. Three-dimensional (3D) crystals of IMPs can sometimes be obtained in the presence of naturally occurring phospholipids. Such is the case of LacY, which resisted 3D crystallization as a purified protein for years, but was easily crystallized after addition of different amounts of phospholipids (Guan et al. 2006). This strongly supports the role of lipids as co-crystalyzing factors, most likely attributed to strong interactions between some lipid species and the protein. Nevertheless, obtaining 2D and 3D crystals of IMPs is still a challenge for researchers.

The reconstitution of IMPs for the purpose of producing 2D crystals is usually performed in proteoliposomes with one phospholipid component. As in the case above, phospholipids with low transition temperatures, such as POPC or DMPC, are used to ensure that the bilayer will be in fluid phase for the purpose of 2D crystallization. Whereas these phospholipids are easy to manipulate in the laboratory, they are far from being biomimetic in structural terms, especially DMPC. As discussed above, natural membranes often contain a high proportion of heteroacid phospholipids, containing a saturated and an unsaturated chain, and often bear charged headgroups. In the case of LacY, the binary biomimetic mixture of the inner membrane of *E. coli*, POPE:POPG (3:1, mol/mol), is always in fluid phase at temperatures above about 28°C. For this reason phase separation is not detected in liposomes by using DSC, unlike in SLBs in the presence of high concentrations of Ca^{2+}, as discussed above. Figure 3.8 illustrates a reconstitution experiment of LacY in preformed SLBs of POPE:POPG. In these experiments, the L_α and L_β phases of SLBs (Fig. 3.7a) are slightly destabilized by addition of a minimal concentration of a surfactant. Afterwards, the purified protein in the same surfactant is added onto the SLBs, followed by an extensive washing process to remove the surfactant with the objective to promote the insertion of the protein into the bilayer (Fig. 3.8b). Medium magnification of the areas with higher corrugation revealed the presence of closely-packed assemblies protruding above the L_α phase (Fig. 3.7c and d). The fluid nature of these domains has been assessed by applying FS and reflects a general behavior of IMPs (Suárez-Germà et al. 2014a).

A question then arises about the exact composition of the fluid phase into which the IMP is inserted, which is relevant for understanding possible lipid-protein selectivity. In this regard, phase diagrams are a valuable tool. Caution should be taken when using phase diagrams obtained from liposomes (Fig. 3.6d) because the substrate (usually mica) induces a shift of the T_m of the SLBs. In the case of POPE:POPG, a pseudo-binary phase diagram for the SLBs has been constructed from AFM topographic images acquired at different temperatures (Suárez-Germà et al. 2014b). The method is based in the calculation of the area covered by the L_α and L_β phases as e.g. POPE:POPG (0.5:0.5, mol/mol) SLBs shown in Fig. 3.9.

From the tie line traced at 27 °C in the pseudo-binary phase diagram (Fig 3.10) an enrichment of POPG in the L_α phases is evidenced, and conversely the L_β phase becomes enriched in POPE. For the biomimetic composition ($\chi_{POPG} = 0.25$), POPG is distributed as follows: $\chi_{POPG}^\alpha \sim 0.79$ (molar fraction of POPG in the L_α phase) and $\chi_{POPG}^\beta \sim 0.11$ (molar fraction of POPG in L_β phase). Depending on particular LPR in the in-plane reconstitution experiments, however, a given amount of phospholipids becomes trapped between the different protein entities. In the case of the experiment illustrated in Fig. 3.8, the analysis of the normalized surface autocorrelation function corresponding to the topological AFM images indicates that each LacY entity would be surrounded by a microdomain consisting of three boundary lipid shells in fluid phase (Picas et al. 2010), that will be highly populated with POPG, according to the calculations performed on the pseudo-binary phase diagram. This, however, does not mean that LacY has a preferential selectivity for POPG; conversely, as will be discussed in Sect. 4.4, the selectivity is actually for the minority component present in the fluid phase, that is POPE.

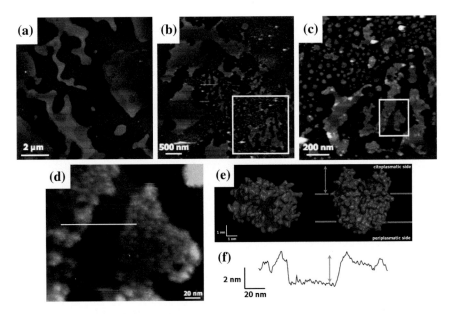

Fig. 3.8 Tapping Mode®-AFM images showing the topography of SLBs of POPE:POPG (3:1, mol/mol) after performing protein incorporation in the absence (**a**) and in the presence (**b**) of 20 μg/ml of LacY, respectively. Progressive magnifications on the highlighted area (*white squares*) (**c**) reveals segregated domains constituted by round-shaped entities that correspond to individual monomeric proteins (**d**). Model of LacY, the frontal cytoplasmic face and lateral view showing the region of the protein embedded in the bilayer (**e**). Height profile along *white line* in (**d**) showing the step-height difference between the protein assemblies and the lipid bilayer in which it is embedded. Adapted from Picas et al. (2010) with permission from Elsevier Science

3.4 Boundary, Non-boundary and Bulk Lipids

Lipids near an IMP should feel its presence through short range interaction forces. These forces were earlier inferred through DSC measurements, which suggested that lipids in close proximity to the IMPs exhibit different physicochemical properties than those in the lipid matrix (Houslay and Stanley 1982). The existence of two populations of lipids in the presence of cytochrome c oxidase reconstituted in liposomes was first reported using electron paramagnetic resonance (EPR) spectroscopy, also known as electron resonance (ESR) spectroscopy (Marsh 1998). In this technique, also used to investigate the structure and dynamics of IMPS (Hubbell and Altenbach 1994), lipids (or specific aminoacids in proteins) are labeled with free radicals with unpaired electros. The most commonly used spin label contains the nitroxide moiety which is very reactive and stable over a wide range of temperatures and pHs. Basically, the analysis of the spectra provided information about the rate of lateral diffusion of the labeled lipid. Thus, the spectra suggested the existence of a lipid population with identical isotropic tumbling equivalent to that seen in absence of the protein; and a second population

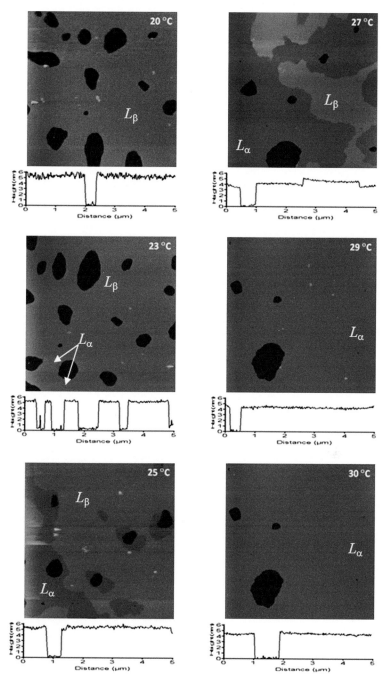

Fig. 3.9 AFM topographic images for POPE:POPG SLB with $\chi_{POPG} = 0.50$ from 20 to 30 °C. *Bottom profile* shows a representative height profile from the corresponding image. Domènech et al. (unpublished work)

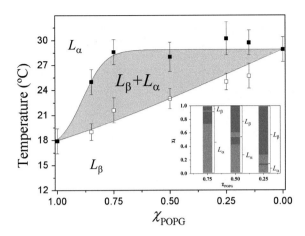

Fig. 3.10 Pseudo-binary phase diagram constructed from analysis of the topographic AFM images of POPE:POPG SLBs, after connecting the corresponding T_{onset} (*open squares*) and T_{offset} (*full squares*). The proportions at 27 °C of each component in each phase (L_α or L_β) for $X_{POPG} = 0.25$, 0.50, and 0.75 obtained after the lever rule application is shown in the *inset*: *blue* for POPE and *green* for POPG

of motionally restricted lipids. This leads to the implication of the existence of boundary lipids around proteins in membranes that have different physical properties than lipids that are further away from the proteins, bulk lipids. The lipids in the first adjacent shell around a protein are sometimes referred to as "boundary lipids" (Fig. 3.11b) where "non-boundary lipids" refers to phospholipids that bind non transiently with high affinity to hydrophobic clefts of the IMPs and often

Fig. 3.11 IMP embedded into the bilayer (**a**), theoretical multi boundary region (*pink*) and bulk lipids (*grey*) around the IMP (**b**) and non-boundary lipid (**c**)

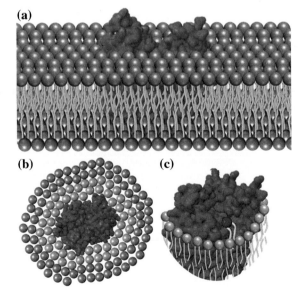

co-crystallize with them (Fig. 3.11c). Such co-crystallization occurs with the potassium channel KcsA and CL of *Streptomyces lividans*. KcsA activity has been found to be dependent on the presence of negatively charged phospholipids, which suggests a direct interaction between these lipids and the protein.

The existence of such a structure with a lifetime in the order of time required for a protein functional event raises the question of how stable such a boundary region could be and whether these boundary lipids might participate somehow in the protein activity.

Noteworthy, in kinetic terms, the rate of exchange between lipids in the bulk of an average fluid membrane bilayer is ~10^{-7} s, an order of magnitude shorter than the average time of 10^{-6} s estimated for most IMP functional events. With such a fast rate of exchange in comparison to the rate of function activity, the bulk lipids can hardly differentially affect IMP behavior. In fact, the existence of a boundary lipid region around IMPs has long been a matter of controversy, mostly because of the reduced residence times of lipids at the protein-lipid interface as compared to residence times in the bulk lipid domains. All in all though, it is convenient to define the boundary region as a lipid shell at the perimeter of the intramembrane protein formed when specific phospholipids bind preferentially to the hydrophobic and/or hydrophilic surfaces of a membrane protein. The residence time of lipids in the boundary region, as well as the lipid-protein stoichiometry, can be determined from EPR experiments performed using spin labeled phospholipids and other techniques. However, we have to consider the time window range for each technique. Thus, while in the case of EPR, the exchange rate of the spins between the annulus and bulk lipids is <10^{-8} s^{-1}; for NMR the exchange rate is >10^{-5} s^{-1}. This means that on the typical time scale of ^{2}H-NMR, the exchange between boundary and bulk lipids cannot be resolved. Hence, most of ^{2}H-NMR studies do not show any immobile lipid component. Accordingly, the residence time at the boundary region should be between 10^{-8} and 10^{-5} s. Actually, the residence time becomes the most adequate parameter for defining the boundary region. Marsh and co-workers (Marsh 1998) have extensively investigated the nature of the boundary region by means of EPR. The stoichiometry of lipid-protein interaction is important, since it may represent the minimum number of lipids required to support a protein activity event. Boundary lipids and its lipid to protein stoichiometry may also be important in sealing the protein-lipid interface in order to avoid membrane leakage. For a membrane containing spin-labelled and unlabeled lipids, a general scheme for the exchange equilibrium between the spin-labelled, L^*, and unlabelled lipids, L, that interact with the protein P, can we written as

$$L_{N_b}P + L^* \Leftrightarrow L_{N_b-1}L^*P + L$$

where N_b is the number of lipid association sites on the protein. By assuming multiple association sites the equilibrium constant can be written as

$$\frac{(1-f)}{f} = \frac{(n_T/N_b - 1)}{K_r} \qquad (3.1)$$

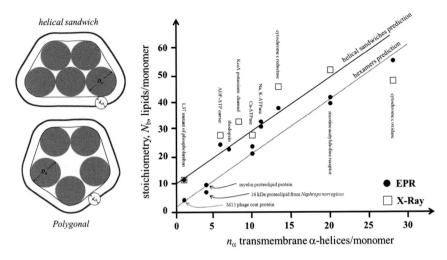

Fig. 3.12 TM α-helices in a helical sandwich (*top left*) and in a regular polygonal arrangement (*bottom left*). The number of acyl phospholipids (N_b) that can be accommodated as a first shell (surrounded highlighter); D_α and d_{ch} are the diameter of the helix and a single acyl chain, respectively (*left*). Dependence of the number of first-shell motionally restricted. The *solid line* is the predicted dependence for monomeric helical sandwiches and the *dashed line* is the corresponding prediction for protein hexamers (*right*). Adapted from Marsh (1998)

where f, is the fraction of spin-labelled lipid that is motionally restricted, K_r is the average association constant of the spin-labelled lipid relative to the host lipid and n_T is the total lipid-protein ratio. Since f is directly obtained from the spectra and n_T is known, N_b and K_r can be determined by using Eq. 3.1 simultaneously under different conditions.

The stoichiometry, on the other hand, can be modelled by straightforward geometrical considerations using two kinds of arrangements, the helical sandwich and the regular polygonal assembly (Fig. 3.12). In these cases, N_b can be obtained from the following equation

$$N_b = \pi(D_\alpha/d_{ch} + 1) + n_\alpha D_\alpha/d_{ch} \qquad (3.2)$$

where n_α is the number of transmembrane α-helices, D_α is the helix diameter and d_{ch} the diameter of the lipid chain. Equation 3.2 is generally valid for multilayer sandwiches in the range $1 < n_\alpha < 7$. For a centered hexagonal arrangement of proteins with $n_a > 7$, modifications with other equations have to be introduced (Marsh 2008). Thus, when the values for N_b are plotted versus the predicted number of transmembrane helices per monomer (n_α) (Fig. 3.12) there is a linear relationship which is maintained for proteins with a single monomeric helix (e.g., phospholamban mutants) and the 7 helix Rho, but deviates for proteolipid hexamers and fails for polytopic proteins with more complex membrane topologies such as cytochrome oxidase.

The type of usefull information obtained from EPR can be illustrated by studies on the Ca^{2+}-ATPase of the sarcoplasmic reticulum. According to EPR measurements, the number of lipids forming an annulus around each protein is 32 at 0 °C, which matches very well with the number obtained from simple geometrical considerations. Thus, assuming that the hydrophobic surface of Ca^{2+} ATPase is 0.14 nm and the molecular area and diameter for a lipid in fluid phase are 0.7 and 0.94 nm^2, respectively, 30 lipid molecules will be required to form a bilayer shell around the protein (Lee 2003). It is interesting to note that, even though DSC is a time independent technique, DSC has been used to estimate the number of lipids present in the boundary region. This number can be obtained by ploting the enthalpy change of the transition as a function of the LPR ratio followed by extrapolation of the usually obtained straight line to zero enthalpy. By using this method it has been estimated that 45 molecules of DMPC form the boundary shell of Ca^{2+}-ATPase.

The controversy on the existence or not of the boundary might find resolution in considering that anular lipids might be better considered as an operational concept than a topographical feature of the membrane. In physical terms, the acyl chains of the lipids at the boundary or boundary region would be expected to undergo significant distortions to adapt to the irregular surface of the protein, thus becoming disordered in comparison with the bulk lipids. If so, the phase transition of phospholipids in the boundary region would decrease in comparison with the transition of the bulk lipids. Unfortunately, this microscopic behavior cannot be detected easily.

3.5 Hydrophobic Match and Mismatch

The mechanism of action of many IMPs implies several changes in their conformation, and therefore requirements for interplay with the surrounding phospholipids must exist to avoid leakage of material through the bilayer. A large amount of experimental evidence supports the existence of hydrophobic matching between lipids and IMPs. This means that the *hydrophobic length* (L_p) of the transmembrane protein domain should match with the *hydrophobic thickness* (L_L) of the lipid bilayer (Fig. 3.13). In turn, this implies that the acyl chains in contact with IMPs should adapt to the irregular surface of the protein by elongation or contraction and/or bending, acquiring a different conformation to that of the bulk lipid, and therefore displaying a different thermotropic behavior, which can be observed by DSC as well as by longer residence times, obtained from EPR experiments, compared with phospholipids in the bulk. These experiments are the basis of the idea that "phospholipids are taken out from the transition" by the IMPs.

The match between L_p and L_L has turned out to be a central feature in lipid-protein interactions and for the regulation of IMPs functions. A hydrophobic mismatch (H.M.) (Jensen and Mouritsen 2004) occurs if the L_P of an IMP does not match with the L_L of the bilayer, and can be defined as

$$H.M. = L_P - L_L \qquad (3.3)$$

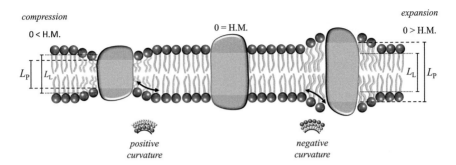

Fig. 3.13 Cartoon illustrating the concept of hydrophobic matching of the lipid species around a protein adapted from Dumas et al. (2000) and Brown (2012). Depending on the size of the hydrophobic region of a protein (L_p) the bilayer thickness (L_L) will contract or expand to best match such a region

By definition, the H.M. is positive if $L_p > L_L$ and that results from bending and stretching of the phospholipid species around an IMP, so that the molecular area occupied by a boundary lipid comes smaller as the lipid becomes closer to the IMP. This behavior (Fig. 3.13 right) is typical of lipids such as PEs that tend to form H_{II} phases. Conversely, the H.M. is negative if $L_p < L_L$, this results from bending and compression of the phospholipid molecules around the IMP. This effect will be exerted by lipids that tend to form H_I phases (see Sect. 4.6). In such cases the area occupied by the lipids near the IMP becomes larger (Fig. 3.13 left) as has been seen in association with glycolipids. Ideal matching occurs when $L_p = L_L$. There are several lines of evidence on the selectivity between IMPs and specific lipids which depend upon lipid-IMP matching. Importantly, the activity of many IMPs has been shown to be sensitive to adequate matching.

An example is provided by the reconstitution of Br in DLPC:DSPC (0.25:0.75, mol/mol) proteoliposomes that have been analyzed by a combination of techniques combining FRET experiments, fluorescence of polarization and computational approaches in light of the binary phase diagram of the lipids (Dumas et al. 1997). In these experiments, at low temperatures where DSPC is in gel phase and DLPC is in fluid phase according the phase diagram of the DLPC:DSPC mixtures, Br was shown to be associated with DLPC. Conversely, at high temperatures, when both phospholipids are in fluid phase, Br prefers DSPC. At low to intermediate temperatures DSPC is laterally segregated into gel domains and can not provide adequate matching to the protein. At higher temperatures, above the upper limit of the melting temperatures of the lipid mixtures, fluid DSPC with lower bilayer thickness than solid DSPC provides an optimum matching environment for Br. At high and low temperatures hydrophobic matching is accomplished because the protein recruits the most adequate phospholipid for its needs. Similar behavior has been observed when LacY was reconstituted in POPE:POPG at different molar ratios and also for Rho reconstituted in different lipid matrices (see Sect. 3.5).

Neglecting the entropy of mixing of lipids and proteins, the Gibbs energy of the system can be reduced to the following expression

$$G = G^o + k\left(\frac{\rho_P}{\pi\xi_L} + 1\right)|L_P - L_L|^2 \tag{3.4}$$

where G^o accounts for the Gibbs energy of the unperturbed bilayer, k is a phenomenological constant related to the compressibility modulus of the bilayer, ξ_L is the persistence length of the lipid bilayer, and ρ_P is the circumference of the protein (assumed to be a cylinder). ξ_L is a measure of the lipid cooperativity underlying the formation of a domain. ξ_L ranges from the single layer of lipids around the IMP to a large number of layers. A commonly used reasonable value of ξ_L is 1 nm (Piknova et al. 1993). The value of ξ_L may be modified by changing the temperature or adding different substances, which may result in a different lateral organization of the proteins in the bilayer. Within this context, a thermodynamic model that contemplates the adaptation of the lipids and the induction of protein segregation when there is a large mismatch between lipids and IMPs (the so-called *mattress model*) was introduced to rationalize intramembrane protein-lipid hydrophobic interactions (Mouritsen and Bloom 1984).

This thermodynamic model describes the phase behavior of the lipid membrane when a bilayer-spanning IMP is embedded within the bilayer where the IMP transmembrane length does not match the thickness of the pure lipid bilayer in equilibrium. It is assumed that the underlying mechanism for IMPs-lipid selectivity and/ or the preference for a particular lipid phase determine the best match. A systematic experimental approach to investigate the H.M. was performed by reconstitution of *Melibiose permease* (MelB), a 12-TMS protein of *E. coli,* in phospholipid matrices differing in acyl chain length and composition (Dumas et al. 2000). In that work, the shifts in transition temperature (ΔT) of the lipid matrices in presence of MelB were fitted to the following equation

$$\Delta T = 16\xi_L^2\left(\frac{\rho_L}{\pi\xi_L} - 1\right)\left[\frac{(L_L - L_P)}{(L_{L,\alpha} - L_{L,\beta})}\right]\chi_P \tag{3.5}$$

where L_L accounts for the average hydrophobic thickness of the bilayer, being $L_{L,\alpha}$ and $L_{L,\beta}$, the hydrophobic thicknesses of the fluid and gel phase, respectively. The authors showed that the T_m shifted linearly with the concentration of the protein (χ_p), and that ΔT is related to the hydrophobic mismatch. The study demonstrates that MelB responds to the hydrophobic matching, with the optimal matching occurring with phospholipid with an L_L of ~3 nm. In this experimental case, DPPC was the best candidate satisfying the hydrophobic matching condition for $L_P = 3.2$ nm. Importantly, it was also demonstrated that the transport activity of MelB is modulated by the acyl chain length of the surrounding lipids. That is, when $L_L \sim L_P$, the activity of MelB is maximal. Another interesting case is Na$^+$, K$^+$-ATPase from the sarcoplasmic reticulum, for which it has been shown that the activity is higher when the enzyme is reconstituted in the saturated DMPC than in the monounsaturated 1,2-dimyristelaidoyl-*sn*-glycero-3-phosphocholine (DMLPC), which can be attributed to the favorable matching provided by DMPC. Thus, for monosaturated PCs, there is a clear dependence of the hydrolytic activity on acyl chain length, the optimum activity is shown when the enzyme is reconstituted in 1,2-dierucoyl-*sn*-glycero-3-phosphocholine (DEPC) (22:1/22:1) carbons with an estimated $L_L = 3.4$ nm. Another demonstration that hydrophobic matching is important in supporting ATPase activity comes from a set of experiments performed using CHOL as a second component of the phospholipid matrix. The addition of 40 mol % of CHOL results in an increase in the enzyme activity embedded in DMLPC or in DOPC. It is known that the incorporation of CHOL into bilayers of fluid lipids such as the two noted, increases the bilayer thickness, resulting in a better hydrophobic matching with the enzyme. There are also cases when the bilayer thickness and activity of the IMPs cannot be correlated, as it is the case with the Na^+-, Mg^{2+}- ATPase from *Acholeplama laideawii*, which seems to prefer short chains, showing the maximum of its activity for acyl chains of 14 carbons. Mithochondrial ATPase shows maximum activity when the acyl chains of the surrounding phospholipids have lengths of 18 carbons.

3.6 Curvature Stress and the Fluid Surface Model

One intriguing question in membrane biology is the wide diversity of lipids, but another one, even more intriguing, is the presence of significant amounts of non-lamellar lipids in natural occurring membranes (Sect. 2.5). Arising form studies on the influence of phospholipids on Rho photoactivity, a model has been proposed that suggest that activity depends, at least in part, on the physical properties of the lipid that influence the lipid's shape and its consequent tendency to create curvature stress in the membrane.

Rho is an IMP of 7 segments present at the disk membrane of the rod cells and it is the actual visual photoreceptor in vertebrates. The key triggering process is the conversion, a millisecond phenomena, of metharhodopsin I (MI) to metharhodopsin II (MII). The MI to MII transition has been carefully studied in a variety of lipid matrices with the conclusion that the Rho photolysis is indeed modulated by the lipid composition. The native retinal rod outer segment (ROS) disk membranes include PCs, PEs, PSs, PIs asymmetrically distributed in the outer and inner leaflet of the bilayer. Remarkably, docosahexaenoic acid (22:6) represents 47 % of the total acyl chain compositios of ROS membranes. Several observations from the flash photolysis activity of Rho pointed to the involvement of properties of the lipids on function beyond those of fluidity and H.M. On one hand these systematic experiments showed that the full native function of Rho was not achieved for any synthetic phospholipid. On the other hand, H.M. seems to influence the formation of MII but it is not crucial when Rho is reconstituted in PC, PE and PS. Importantly, lipids with a tendency to form HII phase favor the MI to MII conversion.

The overall results on flash photolysis studies performed on Rho reconstituted in proteoliposomes of different lipid compositions give support to a proposal for a new membrane model. This model is supported by the continuum theory of elasticity applied to membranes that are conceived as continuous surfaces and has been termed the flexible surface model (FSM) (Brown 2012). In the FSM the transition from L_α to the H_{II} or cubic phases is a transition that allows certain lipids to overcome the geometrical constraints imposed when they form part of the bilayer, causing a curvature stress field of the fluid bilayer. For a bilayer in equilibrium, there is no lateral tension and the curvature stress is zero. For H_{II} phospholipids, the curvature stress is negative, that is oriented towards the water and gives positive curvature. Conversely for H_I phospholipids, the disposition is towards the acyl chains and gives negative curvature. Phenomenological hydrophobic matching (Fig. 3.13) and spontaneous curvature are clearly related. Under conditions where $Lp \gg L_L$, a large curvature stress occurs that induces lateral separation of the IMPs which, in turn, may result in large segregated domains of proteins, as has been experimentally observed.

3.7 Lipid Rafts

The term "raft" has its origins in isolated membrane fractions extracted with cold (4 °C) detergent solutions. It was found that these extracts were rich in sphingolipids and cholesterol. These finding suggested the existence of aggregates in membranes, which consisted of clusters of cholesterol, sphongolipids and other phospholipids. The research on lipid rafts pushed investigations directed to construction of phase diagrams like the one shown in Fig. 3.4. Although there are still controversies on its existence in living cell membranes, the term raft is used to distinguish lipid microdomains that present a reduced lateral mobility and a different composition and order from the rest of the membrane. Actually, the idea of the raft has been already fitted to the current F-MMM (Fig. 1.18) and it is believed to have important roles in signal transduction.

References

Abramson J, Smirnova I, Kasho V, Verner G, Kaback HR, Iwata S. Structure and mechanism of the lactose permease of *Escherichia coli*. Science. 2003;301(5633):610–615.

Bagatolli LA. To see or not to see: lateral organization of biological membranes and fluorescence microscopy. Biochim Biophys Acta. 2006;1758(10):1541–56.

Brown MF. Curvature forces in membrane lipid−protein Interactions. Biochem. 2012;51(49):9782–95.

Costello MJ, Viitanen P, Carrasco N, Foster DL, Kaback HR. Morphology of proteoliposomes reconstituted with purified lac carrier protein from *Escherichia coli*. J Biol Chem. 1984;259(24):15579–86.

De La Serna JB. Perez-Gil J, Simonsen AC, Bagatolli L a. Cholesterol rules: Direct observation of the coexistence of two fluid phases in native pulmonary surfactant membranes at physiological temperatures. J Biol Chem. 2004;279(39):40715–22.

Dumas F, Sperotto MM, Lebrun MC, Tocanne JF, Mouritsen OG. Molecular sorting of lipids by bacteriorhodopsin in dilauroylphosphatidylcholine/distearoylphosphatidylcholine lipid bilayers. Biophys J. 1997;73(4):1940–53.

Dumas F, Tocanne JF, Leblanc G, Lebrun MC. Consequences of hydrophobic mismatch between lipids and melibiose permease on melibiose transport. Biochemistry. 2000;39:4846–54.

Fidorra M, Duelund L, Leidy C, Simonsen AC, Bagatolli LA. Absence of fluid-ordered/fluid-disordered phase coexistence in ceramide/POPC mixtures containing cholesterol. Biophys J. 2006;90(12):4437–51.

Guan L, Smirnova IN, Verner G, Nagamori S, Kaback HR. Manipulating phospholipids for crystallization of a membrane transport protein. Proc Natl Acad Sci U. S. A. 2006;103(6):1723–6.

Hong-wei S, McConnell H. Phase separations in phospholipd membranes. Biochemistry. 1975;14(4):847–54.

Houslay MD, Stanley KK. Dynamics of Biological Membranes. Chischester: John & Wiley Sons; 1982.

Hubbell WL, Altenbach C. Investigation of structure and dynamics in membrane proteins using site-directed spin labeling. Curr Opin Struct Biol. 1994;4(4):566–73.

Jensen M, Mouritsen OG. Lipids do influence protein function—The hydrophobic matching hypothesis revisited. Biochim Biophys Acta—Biomembr. 2004;1666:205–26.

Koynova R, Caffrey M. An index of lipid phase diagrams. 2002;115:107–219.

Lee AG. Lipid-protein interactions in biological membranes: A structural perspective. Biochim. Biophys. Acta—Biomembr. 2003.

Lin HN, Huang CH. Eutectic phase behavior of 1-stearoyl-2-caprylphosphatidylcholine and dimyristoylphosphatidylcholine mixtures. Biochim Biophys Acta—Biomembr. 1988;946(1):178–184.

Marsh D. Structure, dynamics and composition of the lipid-protein interface. Perspectives from spin-labelling. Bba—Rev. Biomembr. 1998;1376:267–296.

Marsh D. Protein modulation of lipids, and vice-versa, in membranes. Biochim Biophys Acta—Biomembr. 2008;1778:1545–75.

Marsh D, Phil D. CRC Hanbook of lipid bilayers. Boca Raton, Florida: CRC Press; 1999.

Merino S, Domènech Ò, Montero MT, Hernández-Borrell J. Atomic force microscopy study of *Escherichia coli* lactose permease proteolipid sheets. Biosens Bioelectron. 2005;20(9):1843–6.

Milhiet P-E, Gubellini F, Berquand A, Dosset P, Rigaud J-L, Le Grimellec C, et al. High-resolution AFM of membrane proteins directly incorporated at high density in planar lipid bilayer. Biophys J. 2006;91(9):3268–75.

Mouritsen OG, Bloom M. Mattress model of lipid-protein interactions in membranes. Biophys J. Elsevier 1984;46(2):141–153. doi:10.1016/S0006-3495(84)84007-2.

Picas L, Carretero-Genevrier A, Montero MT, Vázquez-Ibar JL, Seantier B, Milhiet PE, et al. Preferential insertion of lactose permease in phospholipid domains: AFM observations. Biochim Biophys Acta—Biomembr. Elsevier B.V. 2010;1798(5):1014–1019.

Piknova B, Perochon E, Tocanne JF. Hydrophobic mismatch and long-range protein/lipid interactions in bacteriorhodopsin/phosphatidylcholine vesicles. Eur J Biochem. 1993;218(2):385–96.

Poveda JA, Encinar JA, Fernández AM, Mateo CR, Ferragut JA, González-Ros JM. Segregation of phosphatidic acid-rich domains in reconstituted acetylcholine receptor membranes. Biochemistry. 2002;41(40):12253–12262.

Poveda JA, Fernández AM, Encinar JA, González-Ros JM. Protein-promoted membrane domains. Biochim Biophys Acta—Biomembr. 2008;1778(7–8):1583–1590.

Schmidt ML, Ziani L, Boudreau M, Davis JH. Phase equilibria in DOPC/DPPC: Conversion from gel to subgel in two component mixtures. J Chem Phys 2009;131.

Suárez-Germà C, Domènech Ò, Montero MT, Hernández-Borrell J. Effect of lactose permease presence on the structure and nanomechanics of two-component supported lipid bilayers. Biochim Biophys Acta—Biomembr. 2014a;1838(3):842–52.

Suárez-Germà C, Morros A, Montero MT, Hernández-Borrell J, Domènech Ò. Combined force spectroscopy, AFM and calorimetric studies to reveal the nanostructural organization of bio-mimetic membranes. Chem Phys Lipids. 2014b;183:208–17.

Veatch SL, Keller SL. Seeing spots: Complex phase behavior in simple membranes. Biochim Biophys Acta—Mol Cell Res. 2005;1746:172–85.

Zhao J, Wu J, Heberle FA, Mills TT, Klawitter P, Huang G, et al. Phase studies of model biomembranes: complex behavior of DSPC/DOPC/Cholesterol. Biochim Biophys Acta. 2007;1768(11):2764–76.

Chapter 4
Dependence of Protein Membrane Mechanisms on Specific Physicochemical Lipid Properties

Abstract In Chap. 3 we have shown some examples of how lipid-protein interactions lead to laterally segregated structures in membranes, and how the activity of proteins is related to physical properties of the phospholipids. In this chapter we will first discuss the relationship between membrane structure and bioenergetics, emphasizing that lipids may be part of the machinery involved in proton transport between protein components of the respiratory chain. Second we will present selected examples that relate membrane protein activity with specific phospholipids and we will discuss how this can be rationalized theoretically by introducing the concept of a lateral pressure profile of the membrane. Since the magnitude of lateral pressure within the membrane cannot be experimentally measured, we will show how using atomic force microscopy in force mode and single-molecule force spectroscopy, we can extract nanomechanical properties of the membranes related to protein packing. These properties, in particular the unfolding force or the force required to extract a membrane protein from a bilayer, are related to both the lateral pressure of pure lipid monolayers and the intrinsic surface curvature of monolayers. Finally, we will discuss the application of FRET to identify the phospholipid species present at the lipid-protein interface.

Keywords Lipid coupling · Intrinsic curvature · Lateral pressure profile · Protein activity

4.1 Gibbs Energy and ATP Synthesis: The Lipid Coupling

Physicochemical properties measured in membrane models can provide insight into lipid-protein interactions in membranes and their role in various membrane processes. An example of an active role for phospholipids in membrane chemical reactions emerges from the field of bioenergetics. The paradigmatic example is the coupling between the proton gradient across the membrane and the generation of ATP in a reversible way. In this regard most textbooks use the abstract concept of

© The Author(s) 2016
J.H. Borrell et al., *Membrane Protein—Lipid Interactions: Physics and Chemistry in the Bilayer*, SpringerBriefs in Biochemistry and Molecular Biology,
DOI 10.1007/978-3-319-30277-5_4

"coupling" to refer to the energy transduction between the electrochemical gradient of protons across the biological membranes and the energetic requirement for the ATP synthesis as presented in the chemiosmotic theory (Mitchell 1961). This form of coupling is different what is more conventionally understood to happen between chemical reactions that share a common component.

The main question consists in knowing how the gradient of protons formed during the vectorial transport across the membrane is coupled with the ATPase protein complex. In this regard, coupling coefficients and the degree of coupling can be calculated by using the framework of irreversible thermodynamics formulations by developing the so-called constitutive equations (Caplan and Essig 1983). For example, the group of proteins involved in the respiratory chain is physically separated from the ATPases, so conventional chemical coupling as those observed in solution does not describe the phenomenon occurring in the membrane. Remarkably, the possible involvement of lipids in the overall mechanism of the respiratory chain has not been considered theoretically or experimentally. Emergent ideas, however, conceive of a sophisticated mechanism where lipids may act as actual proton traps through which a surface potential is originated, which, in turn, provides a physical basis for the coupling of the activity of regionally separated IMPs of the system (see Fig. 4.1).

In this section we introduce the basic energetics behind ATP synthesis and proton gradient coupling, employing a hypothetical mechanism that includes the involvement of specific lipids in the process. Thus, the electrochemical potential of an ionic species is given by

$$\tilde{\mu}_i = \left(\mu_i\right)_{\phi=0} + Z_i F \phi \tag{4.1}$$

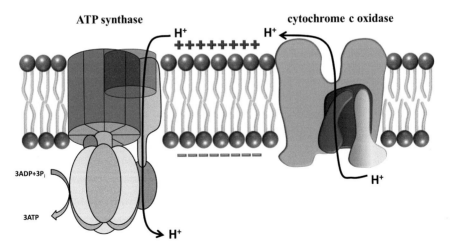

Fig. 4.1 A cartoon showing two of the proteins that conduct oxidative phosphorylation: Cytochrome c oxidase and F_0F_1-ATPase. Headgroups of CL represent a sink for protons, acting as a reservoir and a transceiver (carrier) for the protons required in energy transduction. Based on Haines and Dencher (2002)

where $(\mu_i)_{\phi=0}$ is the chemical potential, Z_i is the charge in terms of the proton charge with the sign of the ion, ϕ the electric potential difference and F the Faraday constant.[1] For the particular case of the proton $Z_i = 1$ and Eq. 4.1 becomes

$$\tilde{\mu}_{H^+} = \mu^{\circ}_{H^+} + 2.3RT \log a_{H^+} + F\phi \tag{4.2}$$

where $\mu^{\circ}_{H^+}$ is the standard chemical potential and a_{H^+} the hydrogen ion activity. When $pH = -\log a_{H^+}$[2] Eq. 4.2 becomes

$$\tilde{\mu}_{H^+} = \mu^{\circ}_{H^+} - 2.3RTpH + F\phi \tag{4.3}$$

If, as it occurs, there is a different concentration of H^+ associated and a different ϕ across a membrane, there will be a transmembrane electrochemical potential difference that we can express as

$$\Delta\tilde{\mu}_{H^+} = F\Delta\phi - 2.3RT\,\Delta pH \tag{4.4}$$

where $\Delta\phi$ and ΔpH are the transmembrane potential and pH differences across the membrane, respectively. Equation 4.4 tells us on how much energy is required or released (depending on the direction of the flow), to transport 1 mol of protons across the membrane. The form most commonly encountered of this equation in biochemistry textbooks, at 25 °C and expressed in millivolts, is

$$\Delta P = \frac{\Delta\tilde{\mu}_{H^+}}{F} = \Delta\phi - 59\Delta pH \tag{4.5}$$

which is known in the field as "proton-motive force". The origin of the electrochemical potential or the proton-motive force resides in different transport processes throughout IMPs in the different membranes.

The importance of Eq. 4.4, or its equivalent 4.5, resides in the fact that $\Delta\phi$ and ΔpH across a membrane and the phosphoanhydride bonds in the ATP molecule are interconvertible forms of the chemical potential energy. For the reaction

$$H^+ + ADP^{3-} + HPO_4^{2-} \Leftrightarrow ATP^{4-} + H_2O$$

the expression of the Gibbs energy change[3] is given by

$$\Delta_r G' = \Delta_r G^{\circ\prime} + RTLn\frac{\left[ATP^{4-}\right]}{\left[ADP^{3-}\right]\left[HPO_4^{2-}\right]} \tag{4.6}$$

---

[1]The Faraday constant is equal to the product of the Avogadro constant and the proton charge: $F = N_A e = 96485.309$ C mol^{-1}.

[2]Concentration, ideally in molal scale, will be used by assuming that the activity coefficients of H^+ and ions in general are nearly the same at both sides of the membrane.

[3]$\Delta_r G'$ is the "transformed" Gibbs energy change, and it refers to the value of the magnitude at a given T, P, pH and ionic strength (I).

where $\Delta G^{o\prime}$ is the Gibbs energy change, used or released, during a chemical reaction under standard conditions when the chemical activities of all the reactants are equal to 1. For the complete coupling reaction the total Gibbs energy change will be zero. Hence it can be written as

$$\Delta_r G'_T = n\Delta\tilde{\mu}_{H+} + \Delta_r G' = 0 \tag{4.7}$$

where n represents the number of protons involved. Equation 4.7 expresses, in thermodynamical terms, the energetic coupling between ATP synthesis (or hydrolysis) and the electrochemical potential.

As noted above, an intriguing question still unsolved is whether phospholipids, most likely located near the proteins of the respiratory chain, could play an active role in molecular mechanims of proton pumping and electrochemical potential establishment via specific lipid-protein interactions. In this regard two IMPs of the electron transport chain of the mitochondria, Cytochrome c oxidase (complex IV) and F_oF_1-ATPase and diphosphatidylclycerol (CL) represent a link between the ATP generation and the proton motive force generation in mitochondria and in aerobic bacteria.

Cytochrome c oxidase is a large transmembrane protein complex, that catalyzes the oxidation of Cytochome c and reduces O_2 to water and generates a $\Delta\tilde{\mu}_{H+} \sim -240$ mV. With the overall reaction being

$$4Cytc^{2+} + 4H^+ + O_2 \rightarrow 4Cytc^{3+} + 2H_2O$$

when a pair of electrons is transported through the Cytochrome c oxidase complex, two protons are transported through the membrane. Importantly, it has been shown using EPR spectroscopy that Cytochrome c oxidase interacts preferably with CL molecules located at the annular region. Although the actual distance between Cytochrome c oxidase and F_oF_1-ATPase has not been established, it is likely to be in the nanometer range. It has been hypothesized that CL may act as a proton trap at the bilayer-water interface, with CL headgroups becoming the source of the protons in the translocation process (Haines and Dencher 2002). The exact mechanism may procede throughout an initial step of proton binding to CL headgroups, followed by lateral diffusion along the membrane interface and, at some point, the transference of the proton to the protein donor or acceptor domains (Fig. 4.1). The role of CL as acceptor or donor of protons is inherent to the acidic property of this phospholipid. Related to this concept of a role for CL, the transport of protons between Cytochrome c oxidase and the ATPase being facilitated by CL is supported by electrical potential and fluorescence measurements carried out in monolayers (Prats et al. 1989) where the lateral conduction of protons at the lipid-water interface was measured.

The model described in Fig. 4.1 is tempting and may be just one example of the lipid-protein chemically based interactions involved in other physiological processes in membranes. There are at least three points to be considered in this particular model: first, the enrichment of CL around the IMPs is likely, given that CL has been shown to co-isolate with the proteins involved in oxidative phosphorylation (i.e. Cytochrome c oxidase); second, there will be a difference between the protons actually bound at the interface and the protons in the bulk solution; and third, the transference of the proton from the CL headgroup to some particular residue in the protein which suggests the existence of non-annular lipids (see Sect. 3.3). The underlying mechanism will involve CL providing the protons that are consumed in the oxidative phosphorylation and $\Delta\phi$ is generated while ΔpH across the membrane decreases. Importantly, proton and by extension ion concentration gradients developed in association with lipids might play a role in other processes such as secondary active transport of substrates though the membrane or in rotation of the bacterial flagellar motor.

4.2   Protein Activity Related to Specific Phospholipids

The D-β-hydroxybutyrate dehydrogenase (BDH) from the inner mitochondrial membrane is one of the best documented examples of lipid-requiring enzymes. The enzyme is exposed to the mitochondrial matrix and catalyzes the conversion of β-OH butyrate to oxacelate in presence of NAD$^+$. BDH requies PC for activity and depends specifically on the PC headgroup. The enzyme is integrated in the inner membrane, and it appears that it consists of a single type of subunit. However, it presents a tetrameric disposition in native membranes and when reconstituted in proteoliposomes. In reconstituted systems this enzyme shows specific selectivity for POPC over other phospholipids and its activity increases if the bilayer is in fluid state. The underlying mechanism of BDH, however, seems to be more related to the specific chemical interaction with the PC headgroup, rather than to intrinsic curvatures or hydrophobic matching in the bilayer.

A large body of experimental evidence about the dependence of protein activity on lipid composition has emerged from works where the lipid composition of bacterial membranes has been genetically modified. Using molecular genetics techniques, investigators have manipulated lipid composition to study lipid-protein intereactions in living cells (Dowhan et al. 2004). For example, the mutation of *pssA* or *psd* results in a decrease in the PE levels or an increase in the amount of PS. Among other results, a consequence of the reduction of PE from ~75 to ~35 % in *E. coli* cells is the appearance of filamentous forms and the eventual cessation of growth. These findings seem to be related to an alteration of the cell division machinery due to the decrease in PE with a concomitant increase in the negative surface charge density, because the residual negatively charged lipids make up a higher proportion of the membrane.

Another example of the influence of lipids on protein activity comes from studies on the reconstitution of Lac Y in proteoliposomes. As discussed in Chap. 1, PE acts as a chaperone in the proper folding of LacY. It has been postulated for some time that the *in vivo* function of LacY requires an amino group in the headgroup of the surrounding phospholipids, either PE or PS. This hypothesis assumed that the hydrogen bonding would play a defined role in the LacY-lipid interaction.

Nevertheless, recent *in vitro* studies carried out with proteoliposomes formed with lipid extracts from cells or synthetic phospholipids, point to a more subtle interaction based on the balance of the headgroup and the acyl chains (Vitrac et al. 2013). In these experiments LacY was reconstituted in proteoliposomes of different lipid compositions and its activity was assessed through the uptake of [^{14}C] lactose. When LacY was reconstituted in the total lipid extract from PE-containing cells, the transport of the substrate was at a maximum. Conversely, when using PE-deficient cells, transport was not observed (Fig. 4.2). These findings support the original idea that the PE headgroup is required for LacY function. However when LacY was reconstituted in lipid extracts where PE had been replaced by PC, active transport was also significant which is in agreement with observations made *in vivo* where cells with 70 % of PE or PC both have shown, similar levels of transport. These later observations would be consistent with an explanation of the findings from genetically manipulating of lipid amounts in cells that are noted above, where a surface charge density could be important for activity. Both PE and PC would likely maintain a similar surface charge density in the membrane.

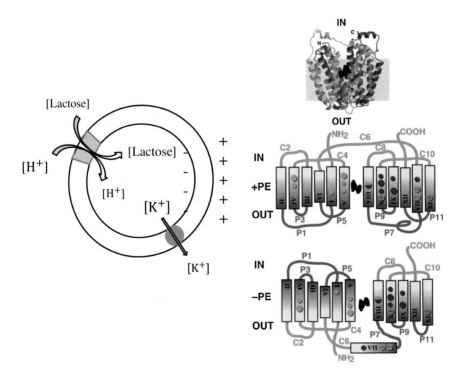

Fig. 4.2 Scheme of LacY permease activity assay. Proteoliposomes of LacY are formed in 50 mM KPi and diluted 200-fold in 50 mM NaPi containing radiolabeled lactose in presence of valinoycin to generate $\Delta\Phi$ across the membrane. *Uphill* transport is driven when $[K^+]_{in} > [K^+]_{out}$ at the time of the dilution. The loss of transport ability correlates with misfolding of periplasmic domain P7 that is crucial for the active transport of substrate. Reprinted from Dowhan et al. (2004) with permission from Elsevier Science

To get new insights on the possible involvement of the acyl chains of the phospholipids on LacY transport, the protein was reconstituted in matrices constituted by synthetic phospholipids (containing homo- and heteroacids) at the molar ratio found in the inner membrane of *E. coli*. The outcome was that the heteroacid phospholid POPE was the lipid that provided the highest activity to LacY, followed by POPC, DOPE and DOPC. The main conclusion arising from these studies was that heteroacid phospholipids, with one saturated and one unsaturated acyl chain, provide the best molecular environment (plasticity) for active transport. This body of results strongly suggests that, at least for LacY, both the headgroup and the specific acyl chain composition, participate in the overall interplay of the lipid with the protein. Since they find their origins in the lipid structural properties, it is likely that physicochemical properties such as intrinsic curvature of the adjacent lipids or their electrical nature would determine the interactions of lipids with most IMPs.

The remarkable influence of phospholipids has been observed in voltage-regulated channels that open and close in response to changes in the transmembrane potential. Ion channels are of great relevance because they are involved in many cellular functions including the propagation of electrical signals that innerve axons and nerve terminals or muscle. The group of voltage dependent cation channels includes, members with specificity for K^+, Na^+ and Ca^{2+} ions. Interestingly, all voltage-gated ion channel proteins are related in structure and function. One of the most investigated is the voltatge-dependent KvAP K^+ channel from the archaebacterium *Aeropyrum pernix* (Jiang et al. 2003). The importance of the composition of the lipid membrane in the gating mechanism has been evidenced by electrophysiological measurements performed on the KvAP K^+ channel reconstituted in phospholipid matrices (Schmidt et al. 2006). Whilst KvAP reconstituted in POPE:POPG shows voltage dependent currents, when reconstituted in DOTAP (positively charged) the channel does not open. One of the conclusions of these studies was that the presence of an anionic group, usually a phosphate group, appears to be crucial for KvAP function. As opposed to the case of LacY, where charge density on the lipids appears to be a strong modulator of activity, in the case of KvAP, the charge sign on the lipids appears to be the important factor, suggesting that the effect on the protein activity might have more chemical than physical origins.

Another example of a K^+ channel that is dependent on membrane phospholipids is the KcsA channel from *Streptomyces lividans,* which has a different architecture than the KvAP channels. The KcsA channel has a homotetramer structure with four identical protein subunits that form a cone with a central pore, each subunit of KcsA containing two TM helices. The base of the cone faces the extracellular phase and the apex is at the internal face of the membrane. The channel consists of an inner pore facing the cytoplasmic phase, a large cavity near the middle of the pore and the selectivity filter that separates the cavity from the extracellular phase (Fig. 4.3). The KscA K^+ channel opens at acidic pHs and, notably, is only active in the presence of anionic lipids. The role of the phospholipids can be assessed by measuring $^{86}Rb^+$ flux uptake in proteoliposomes of KcsA made with POPE:POPG (3:1, mol/mol) and POPE:POPC (3:1, mol/mol). It has been demonstrated that the channel only opens when PG or another anionic phospholipid such as PS or CL is

Fig. 4.3 Side view (**a**) and extracellular view (**b**) of KcsA showing the locations of tryptophan residues (W67, W68, W87). The bound lipid DAG is shown in space-filling and ball-and-stick formats K$^+$ ions are shown in space-filling format. Reprinted from Marius et al. (2005)

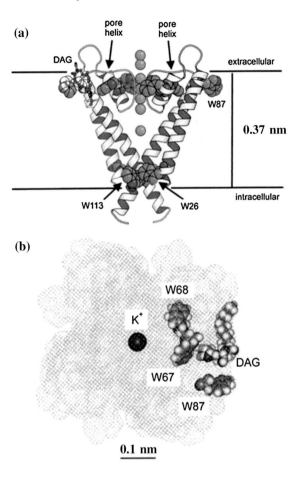

present. The requirement for PG seems related to the ion conduction through the channel. It is also interesting to note that PG is copurified along with KcsA and that it appears in the crystal structure filling a groove on the surface between adjacent units (Valiyaveetil et al. 2002). It is believed that the negatively charged headgroup of the PG molecule may interact with Arg-64 and Arg-89 (Fig. 4.3a), resulting in the influence of non-annular anion lipid binding on the activity of the KcsA channel (Weingarth et al. 2013).

In order to unveil the selectivity of binding sites of IMPs, fluorescence quenching methods are often applied. In KcsA, the quenching of engineered tryptophan residues by brominated phospholipids has been used (Marius et al. 2005). In these studies the fluorescence quenching was described using

$$I = I_{min}^A + \left(I_o - I_{min}^A\right)\left(1 - f_{Br}^A\right)^n \tag{4.9}$$

where I_o and I_{min}^A are the fluorescence intensities of KcsA in the presence of non-brominated and brominated lipids, respectively, and n, the number of annular lipid

Table 4.1 Lipid binding constants for W26 determined from fluorescence quenching plots

Lipid	Annular sites on extracellular side relative binding constant	Nonannular sites Binding constant (mol fraction^{-1})	Annular sites on intracellular side relative binding constant
PA	0.80 ± 0.17	4.6 ± 2.1	2.93 ± 0.77
PG	0.68 ± 0.07	3.0 ± 0.7	1.88 ± 0.10
PS	0.94 ± 0.11	7.1 ± 2.7	0.70 ± 0.30
CL	0.77 ± 0.10	7.3 ± 1.2	0.70 ± 0.16

Reprinted from Marius et al. (2005) with permission from Elsevier Science

binding sites; and, f_{Br}^A is the fraction of annular sites on KcsA occupied by the brominated lipids which is given by

$$f_{Br}^A = \frac{K_A \chi_{Br}}{[K_A \chi_{Br} + (1 - \chi_{Br})]} \tag{4.10}$$

where K_A is the binding constant and χ_{Br} is the mole fraction of brominated lipids in the lipid mixture. In the Table 4.1 one can see values of n obtained by using this method. In this particular study it was shown that the binding constant for PS, and CL are close to that of PC and the binding of PA and PG was approximately three- and two-fold stronger, respectively, than PC. In addition, the ion binding to the extracellular side of KcsA is specific.

As discussed before, there is a general agreement on the fact that IMPs (except non-lipid anchored proteins) prefer fluid phases over ordered phases. An experiment that illustrates this behavior has been carried out using AFM imaging to investigate the distribution of KcsA in SLBs of POPE:POPG (3:1, mol/mol). (Seeger et al. 2009) (Fig. 4.4). The black line in the image at a temperature of

Fig. 4.4 Temperature-controlled AFM experiment of a SLB of POPE:POPG 3:1 plus KcsA. The imaged area had a size of 100 μm^2. (Generous gift of Dr. Andrea Alessandrini, University of Modena)

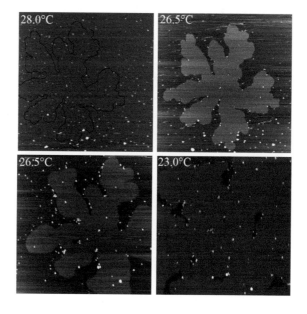

28.0 °C represents the interface of the L_β domain, which developed at lower temperature and is visible in the upper image taken at 26.5 °C. Two images at temperature 26.5 °C are shown since they were allowed for equilibration. Upon the formation of a L_β domain, KcsA was excluded from this domain and it accumulated at the domain interface as it started to aggregate in the L_α phase. Since the fluid phase is enriched in POPG and the gel domains are enriched in POPE (Fig. 3.10) KcsA experiences changes in the local lipid environment. In addition, the physics of the lipid bilayer changes in the phase transition regime.

4.3 Dependence of Protein Activity on Lipid Packing, Order Parameter and Temperature

At this point, it is interesting to know how the phase transition of the bilayer is affected by the presence and activitites of IMPs. A study performed with LacY reconstituted in proteoliposomes of POPE:POPG that combined ATR-FTIR spectroscopy with measurements of protein activity provided insights into the protein-lipid bilayer interplay (le Coutre et al. 1997). In these studies, the effects of varying the LPR were carefully scrutined by correlating the orientation of the protein in the lipid bilayer (namely the average α-helical tilt angle relative to the bilayer normal), the activity of Lac Y assessed by counterflow measurements and the lipid order parameter (S). The values of S are obtained experimentally from the ratio of integrated absorbances from the ATR instrument and the electrical field amplitudes on the germanium lipid interface (Goormaghtigh et al. 1999). As shown in Fig. 4.5, the activity of the protein decreases when the LPR decreases below 800 with a concomitant increase of the average α-helical tilt angle and a

Fig. 4.5 Dependence of lipid order parameter, the average tilt angle and counterflow activity on the lipid-to-protein ratio. Reproduced from le Coutre et al. (1997)

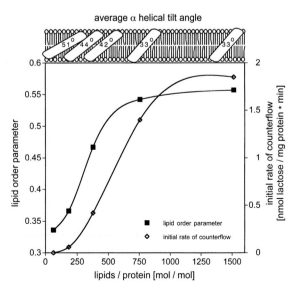

decrease of S. These experiments point to the relevance of the lateral pressure exerted by the lipid bilayer on the protein activity, and in particular the critical role that POPE may play in providing adequate packing within the membrane (le Coutre et al. 1998).

In the presence of an adequate LPR, it is interesting to investigate how temperature affects the viscosity of the bilayer. In this regard fluorescence anisotropy measurements (see Sect. 2.6.2) reported from fluorescence labels as TMA-DPH and DPH allow a direct interpretation. The temperature dependence of polarization of TMA-DPH and DPH of LacY in proteoliposomes reconstituted in POPE:POPG (3:1, mol/mol) and the lipid extract of *E. coli* are shown in Fig. 4.6. As can be seen, the features of the phospholipid phase transition, for TMA-DPH (Fig. 4.6a, b) and DPH (Fig. 4.6c, d) liposomes, were almost unaffected by the presence of LacY. In turn, the anisotropy of TMA-DPH for the POPE:POPG system dropped from 0.34 to 0.24, approximately, when the temperature increased from 0 to 35 °C (Fig. 4.6a), while that of DPH dropped from 0.32 to 0.13 for the same temperature range (Fig. 4.6b). Notably, in this binary lipid system, LacY caused a slight increase in T_m of 0.3 °C as seen by TMA-DPH and it induced a decrease of 1.6 °C in the T_m when DPH was the probe used. This difference might reflect the slightly different regional environments (more or less hydrophobic) of the probes in the bilayer.

When LacY was reconstituted in the *E. coli* lipid extract, TMA-DPH reported lower anisotropy values in liposomes than in proteoliposomes (Fig. 4.6b) and the T_m shifted slightly to a higher value. This reflected an increase in the molecular order at the interface region due to the presence of the protein. When using DPH, the protein also induced a significant effect on the *E. coli* membranes (Fig. 4.6d) by, increasing and decreasing the anisotropy values below and above the main phospholipid phase transition, respectively. Measurements from DPH indicate that the presence of LacY induces an increase of approximately 2 °C in the transition temperature for the *E. coli* lipid matrix. The influences of an IMP on membrane anisotropy are modest as seen in Fig. 4.6 but if we assume that T_m might respond linearly with increasing concentrations of the protein, the decrease of T_m calculated from DPH anisotropy measurements reflects a qualitative change in the phospholipid acyl chain order in the presence of LacY. In addition, the increase of T_m in TMA-DPH liposomes in the presence of LacY suggests that some specific interaction between the headgroups of the phospholipids and LacY are occuring.

Irrespectively of the fluorescent label used, an increase in the temperature results in a decrease of the fluidity of the membrane as measured by the anisotropy values, and that is accompanied by an increase in general IMP activity. The activity can be described using the Arrhenius equation by plotting the IMP activity in terms of substrate uptake as a function of the inverse of the temperature,

$$\ln k = -\frac{E_a}{RT} + LnA \qquad (4.11)$$

where k is formally the rate constant for the reaction, E_a is the activation energy and A the pre-exponential constant term. E_a and κ can be calculated from the slope

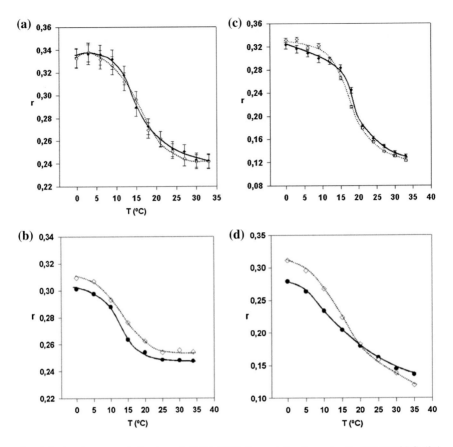

Fig. 4.6 Steady-state polarization of: TMA-DPH for vesicles formed with POPE:POPG (3:1, mol/mol) liposomes (•) and proteoliposomes of LacY (◇) (**a**); TMA-DPH for liposomes formed with the lipid extract from *E. coli* (•) and proteoliposomes resconstituted in the *E. coli* extract (◇) (**b**); DPH for vesicles formed with POPE:POPG (3:1, mol/mol) liposomes (•) and proteoliposomes of LacY (◇) (**c**); and, DPH for liposomes formed with the lipid extract from *E. coli* (•) and proteoliposomes rescosntituted in the *E. coli* extract (◇) (**d**). Merino-Montero (2005)

and y-intercept, respectively, of the plot of the logarithm of the reaction rate versus the reciprocal of the absolute temperature. The Arrhenius plot of the activity of many IMPs shows two straight lines which intercept at what is commonly known as the break point. This can be seen in the Fig. 4.7 where the activity of lactose uptake by right-side-out (RSO) vesicles containing LacY in *E. coli* membranes (usually obtained by mechanical disruption of bacteria after applying a lysozyme) are plotted versus the reciprocal of *T*.

In these experiments, RSO vesicles carrying LacY exhibit a nonlinear Arrhenius plot between 13 and 25 °C with an abrupt discontinuity at ~18 °C, which compares well with the T_m of the *E. coli* lipid extract obtained from the

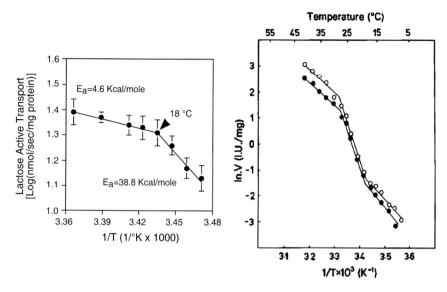

Fig. 4.7 Arrhenius plot of active lactose transport by RSO vesicles from *E. coli* containing LacY (*left*) and Ca^{2+}-ATPase and sarcoplasmatic reticulum (*right*). Reprinted with permission Zhang et al. (2000) © 1997 and Madden and Quinn (1979) © 1997 American Chemical Society

maximum inflexion point of the anisotropy curves in Fig. 4.6b, d. The activation energy at temperatures below the phase transition is about 463 kJ mol^{-1}, while above the T_m, it is 19.3 kJ mol^{-1}. In this case, the break in the Arrhenius plot is likely due to the phase segregation undergone by the *E. coli* lipid extract used to constitute the bilayer. In other cases such as the native Ca^{2+}-ATPase from the sarcoplasmatic reticulum two distinct breaks are observed in the Arrhenius plots at 21 and 15 °C that are proposed to correspond to the uncoupled transitions occurring in both halves of the bilayer. When the Ca^{2+} ATPase is reconstituted in DPPC two break points are also seen in the Arrhenius plot; one occurring at 42 °C which coincides with the T_m of DPPC in non-annular lipid, and a second break at 30 °C, which is attributed to the transition undergone by the lipid molecules at the annular region (Houslay and Stanley 1982). There are other factors that might induce the observed behavior of the Arrhenius plots, such as phase changes of the solvent or conformational changes undergone by the IMPs during their activity. But it seems most likely that changes in the lipid environment of the embedded protein are the source of the observed properties.

4.4 Interactions of Lung Surfactant Through Lipid Monolayers

The interaction of SP-B on binary monolayers of DPPC:DPPG (7:3, w/w) can be inferred from the compression isotherms shown in Fig. 4.8a. As can be seen, at low surface pressure the presence of SP-B induces the expansion of the monolayer to larger molecular areas. This expansion is interpreted as a result of SP-B insertion into the phospholipid monolayer. The monolayers with SP-B inserted show an inflexion point close to 60 mN m^{-1}, which is not observed in the case of protein-free DPPC:DPPG monolayers. This subtle change in the curve is interpreted as the result of the exclusion of SP-B from the monolayer, likely accompanied with some, possibly selective, loss of phospholipid molecules from the interface.

As shown in Fig. 4.8b, similarly to SP-B, SP-C induces an increase in the molecular area of the monolayer after insertion between the phospholipid molecules. As also seen with SP-B, the exclusion of SP-C (and other phospholipid molecules) from the lipid interface is also observed, although in this case at a surface pressure of ~50 mN m^{-1}

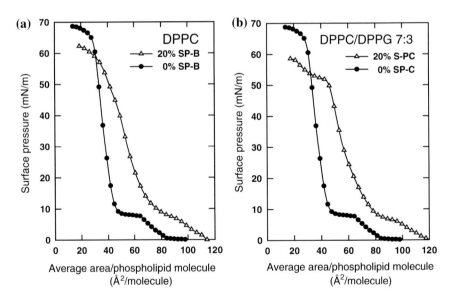

Fig. 4.8 Compression isotherm of the DPPC pure monolayer and DPPC with 20 % of SP-B (**a**); and the monolayer of DPPC/DPPG (7:3, mol/mol) and the same monolayers with 20 % of SP-B (**b**) (Generous gift from Dr. Antonio Cruz, Universidad Complutense of Madrid)

4.5   The Lipid-Protein Interface

Whilst a large amount of evidence indicates a preference of IMPs for fluid phases, it is more difficult to demonstrate how physical properties such as lateral pressure or intrinsic curvature of phospholipids influence IMPs insertion, correct folding and activity. Since the internal lateral pressure cannot be directly assessed either in liposomes or in SLBs, lipid monolayers constitute and indirect but conveni-ent model to provide a means of understanding the selectivity between IMPs and specific phospholipid species. To determine the influence of phospholipid acyl chains on IMPs one could use a strategy of investigating the lateral compress-ibility of monolayers containing IMPs and binary mixtures of lipids with varying chain lengths and amounts of unsaturation. Studies on compression isotherms of POPE, DPPE and POPG, main components of the *E. coli* inner membrane, are of relevance. The surface pressure-area (π-A) compression isotherms for these phospholipids at 37 °C are shown in Fig. 4.9 above along with compressability (C_s^{-1}) values (insets). As expected DPPE, with both acyl chains saturated, remains in a more compressed structure while POPE, with a double bond in the *sn*-2 acyl chain that introduces an intermolecular steric effect, results in an increase in the distances between individual molecules. Surface pressure-area (π-A) compres-sion isotherms for the mixed systems POPE:POPG and DPPE:POPG are shown in Fig. 4.9 below. At a surface pressure of 30 mN m^{-1}, a pressure considered as representative of a bilayer, all the mixed monolayers showed a lower area per

Fig. 4.9 Surface pressure-area isotherms of: DPPE (■), POPE (▲), POPG (○) (*above*) and DPPE:POPG (3:1, mol/mol) (■), POPE:POPG (3:1, mol/mol) (●) 37 °C (*below*). Insets show the elastic moduli of area compressibility (C_s^{-1}) corresponding to each isotherm

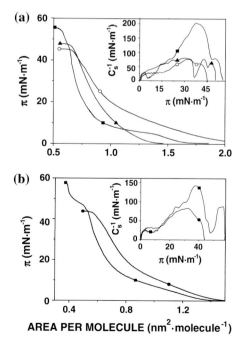

molecule than their corresponding pure phospholipid components. These findings are consistent with a condensing interaction between the individual components. The C_s^{-1} values plotted in the insets of Fig. 4.9 indicate that DPPE:POPG is in a more condensed state than POPE:POPG. IMPs that are inserted in these different lipid matrices might be expected to be differently influenced by the lipid characteristics and the degree of interaction between the lipids.

To get insight into the interactions occurring at the lipid-protein interface, AFM and AFM in FS mode can be used in a new approach to get topographic and nanomechanical information of reconstituted systems. In this approach the AFM cantilever tip is pressed against the PLSs. When the tip and the sample are brought together, ideally, a single IMP attaches by unspecific adsorption to the tip. As the distance between the tip and the PLS is subsequently increased, the IMP is extracted from the bilayer and generates a restoring force that causes the cantilever to bend (Fig. 4.10a). This AFM operating mode is known as AFM-based single-molecule force spectroscopy (SMFS).

This approach has been applied to characterize the influence of the lipid environment on the LacY structure (Serdiuk et al. 2014, 2015). In these experiments LacY was unfolded from POPE:POPG (3:1, mol/mol) and POPG PLSs,

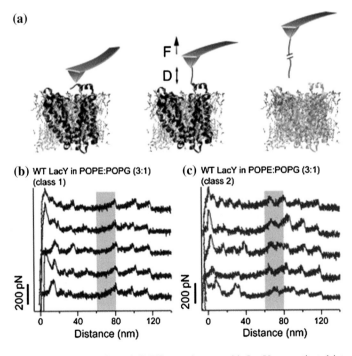

Fig. 4.10 Schematic representation of SMFS experiments with LacY reconstituted into SLB (**a**). Typical force-distance curves corresponding to the unfolding LacY from POPE:POPG (3:1, mol/mol) bilayers. Reprinted from Serdiuk et al. (2015)

yielding two kinds of force-distance curves (Fig. 4.10b, c) which are attributed to the unfolding of the native and inverted topological structures of LacY (Fig. 4.2). Reconstitution into PEs differing in acyl chain composition will provide insight on the role of the lateral surface packing.

SMFS can be used to investigate the mechanical unfolding pathways and structural stability of IMPs (Muller 2008). By reconstituting a protein in SLBs of different composition, it is possible to apply SFMS to determine if different lipids modify the unfolding process. An example of such experiments is shown in Fig. 4.11. As the unspecific interaction between the tip and LacY may occur in any of the loops of the protein, the various force curves represent all possible events, from extraction of a single TM helix to the complete unfolding of the IMP.

The behaviour of a polymer under mechanical stress can be described by the worm-like-chain (WLC) model for elasticity. The force-extension curve, $F(x)$ versus (x), is described by the equation

$$F(x) = \frac{k_B T}{p}\left[\frac{1}{4}\left(1 - \frac{x}{L_{cl}}\right)^{-2} - \frac{1}{4} + \frac{x}{L_{cl}}\right] \tag{4.11}$$

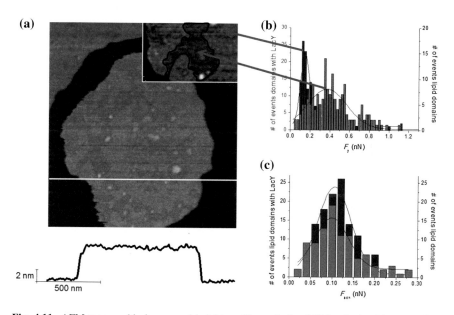

Fig. 4.11 AFM topographic image and height profile analysis of PLSs obtained by spreading LacY reconstituted in proteoliposomes of POPE:POPG (3:1, mol/mol) (Z scale = 15 nm) (**a**) Insert in **a** shows a magnified image (470 × 280 nm, Z = 3 nm) where domains with LacY can be distinguished from domains without LacY (highlighted). Distribution of threshold (F_Y) (**b**) and adhesion (F_{adh}) (**c**) forces of higher and lower domains in (**a**). Reprinted from Suárez-Germà et al. (2014) with permission from Elsevier Science

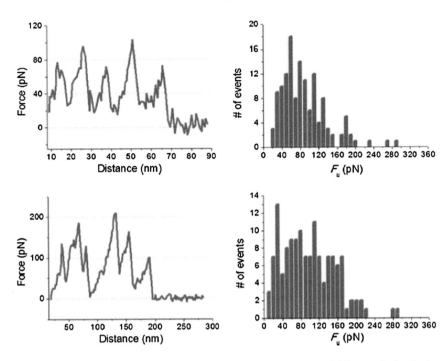

Fig. 4.12 Representative force-distance curve of single LacY unfolding and distribution of unfolding force (F_u)(B) from POPE:POPG (3:1, mol/mol) (*up*) and DPPE:POPG (*down*). Reprinted from Suárez-Germà et al. (2014) with permission from Elsevier Science

where p is the persistence length, which reflects the polymer flexibility and L_{cl} is the contour length. By fitting the worm-like-chain (WLC) model to the peaks observed in the force-distance (*FD*) curves, it is possible to determine the force required to unfold a protein segment (unfolding force, F_u) and also the approximate number of amino acids that the segment contains (Fig. 4.12). The average unfolding rupture forces corresponding to LacY embedded in POPE:POPG and DPPE:POPG are 72.7 ± 3.6 pN and 91.4 ± 4.3 pN, respectively (Suárez-Germà et al. 2014). Therefore the forces governing the protein-lipid interaction indicate that LacY is more tightly inserted in DPPE:POPG than in POPE:POPG. In terms of lateral pressure, the monolayer of DPPE:POPG is more packed and less compressible (greater internal pressure) than the one of POPE:POPG (see Fig. 4.9) and those properties are likely to be the source of the differences observed in the F_u values.

4.6 Thermodynamic Framework and Nanomechanics of IMP Activity

The experimental approach based on AFM topography and SFMS provides experimental evidence on the influence of the lateral pressure and the intrinsic surface curvature on IMP packing and, by extension, on protein function. Understanding how the magnitudes of these characteristics are related to defined conformational states and the transport activity of IMP, however, requires some theoretical discussion. One model of how IMPs sense the surrounding lipid environment is based on the idea that the lateral pressure along the bilayer (π) axis, $p(z)dz$, is not homogeneous as it is observed from compression isotherms of monolayers but includes different components at the nanoscale level (Cantor 1997). This seminal thermodynamic analysis invoked the lateral pressure, specifically the lateral pressure density, as the basic factor involved in the conformational changes undergone by an IMP during its activity. The chemical potential of a protein in a membrane can be written as

$$\mu_b = \mu_b^\circ + k_b TLn(X_b) + \int A_p(z)p(z)dz \qquad (4.13)$$

where $A_p(z)$ is the cross-sectional area of the protein at a distance z from the mid-plane and $p(z)$ is the lateral pressure density, expressed as

$$p(z) = \frac{\partial \pi(z)}{\partial z} \qquad (4.14)$$

where $\pi(z)$ is the lateral pressure acting within a slice of bilayer of thickness δz,

The lateral pressure profile, $p(z)$, is constituted by three components: two repulsive and one cohesive (Fig. 4.13). The first repulsive component has its origin in the electrostatic interactions occurring at the headgroup region (π_{HG}), and the second arising from the steric interactions between the acyl chains (π_{CH}). In the headgroup, the lateral positive pressure arises from steric, hydration and electrostatic interactions, which are normally repulsive but sometimes attractive, such

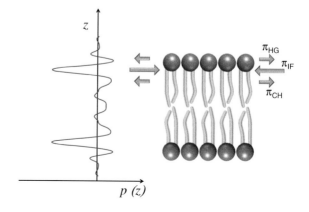

Fig. 4.13 Theoretical lateral pressure profile $p(z)$ with distance z from the bilayer mid-plane in a lipid membrane of DPPC

as when hydrogen bonds occur. The interactions in the acyl chain region are also repulsive and result from a balance between attractive van der Waals interactions and thermal fluctuations of the chains. The third component, a cohesive component, is the pressure localised in the polar-apolar interface between the headgroups and the acyl chains (π_{IF}). The line tension at the interface originates from the unfavorable cost of contact between the acyl chains with water (hydrophobic effect) (Sect. 2.5.1). A tight packing in this region occurs in order to avoid contact between acyl chains and water with the consequence of producing a negative lateral pressure that tends to contract the bilayer. In equilibrium there is a counterbalance beween hydrophobic and repulsive forces and hence there is no tension in the membrane. In other words, the lipid bilayer is then at its Gibbs energy minimum.

This theoretical analysis formally accounts for the phenomenological concept of "plasticity" in lipid-protein interaction that emerged from observations on LacY activity in cells (Bogdanov et al. 2010). It supports implicitly the involvement of the spontaneous intrinsic curvature (c_o) of a monolayer

$$c_o = \frac{1}{r_o} \tag{4.15}$$

with r_o being the spontaneous radius of curvature. And it leads to an alternative understanding of the structures formed by lipids in solution (Sect. 2.5).

Some attempts have been made to measure the lateral pressure profile experimentally by using pyrene-labelled DOPC, with the label located at different positions on the acyl chains (Templer et al. 1998). Acknowledging that the presence of pyrene introduces some molecular distortion, the technique only allows for the measurement of only relative pressures, since exact position of the label in the bilayer cannot be measured. Alternatively lateral pressure profiles can be calculated by using atomistic molecular dynamic simulations (Ollila 2010).

It has been proposed that an IMP "feels" the1999 mechanical stress of the membrane, hence changes in the lateral pressure profile may affect the conformational equilibrium occurring during its activity (Cantor). There are two paradigmatic cases where intrinsic curvature is involved in protein activity: phosphocholine cytidylyltransferase (CTP) and mechanosensitve channels (MscL).

CTP is a rate limiting enzyme that participates in PC biosynthesis and is activated when it binds to bilayers. In studies on CTP activity in lipid bilayers of DMPC:DOPC and DOPC:DOPE, increases in the enzyme activity were observed when the amounts of DOPC were increased, and decreases when H_I amphiphiles (such as lyso-MC) were incorporated into DOPE:DOPC (Attard et al. 2000). Both phenomena were coherently interpreted as a result of the tendency to adopt a more negative curvature when adding DOPC to DMPC or DOPE to DOPC, and to assume positive curvature when adding lyso-PC to DOPE:DOPC.

It has been reported that the large conductance mechanosensitive MscL is strongly dependent on the lateral pressure within the bilayer. Experimentally, the MscL opens when lyso-PC is added to one of the leaflets of the bilayer, which can intuitively be related with changes in the lateral pressure profile (Fig. 4.14). Since

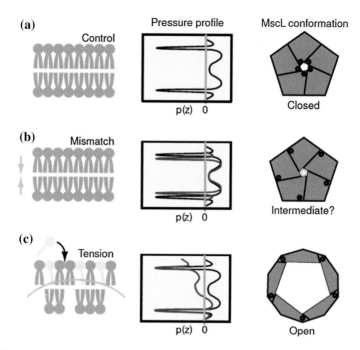

Fig. 4.14   a An unperturbed (control) bilayer stabilizes the MscL in its closed conformation. **b** Reconstitution of MscL into bilayers of different thickness compresses/expands the pressure profile and biases the threshold of activation through mismatch of the hydrophobic regions of the lipid and the protein, possibly stabilizing an intermediate conformation of the channel. **c** Asymmetric incorporation of cone-shaped lipids (i.e. LPC) alters the pressure profile, favoring the fully open state. Reprinted from Perozo et al. (2003) with permission from Elsevier Science

Lyso-PC is prone to form or induce the H_I phase, its insertion into the outer leaflet of the bilayer promotes a positive curvature of the bilayer which results in the opening of the MscL channel. The mechanism behind this conformational change of the protein results from subtle changes occurring in the lateral pressure induced by the incorporation of Lyso-PC in the outer leaflet of the bilayer. The curvature of the bilayer that provokes changes in the lateral surface pressure occurs because of the asymmetry between both leaflets. Conversely, if the bilayer is symmetric, MscL remains closed.

We have discussed how the complexity of the lipid matrix can affect the activity of MscL. There is another influence of PE on MscL channel opening that comes from the capacity of the PE headgroup to form hydrogen bonds with the protein (Elmore and Dougherty 2003).

A quantitative approach has been applied to MscL (Gullingsrud and Schulten 2004), which consisted of calculating the difference in energy between the two conformational states of the protein (non-tilted and tilted states) modeled as a truncated cone (with radius $r(z)$ and slope s) interacting with a lateral pressure profile $p(z)$. Thus, one can write

$$\Delta w = w_{non-tilted} - w_{tilted} = 2\pi R s \int d(z)p(z)z + \pi s^2 \int dz\, z^2 p(z)$$

$$= 2\pi RsM_I + \pi s^2 M_{II} \tag{4.16}$$

where M_I and M_{II} are the first and second moment of the lateral pressure profile, both related to physical properties of the bilayers. We indentify

$$M_I = k_o c_o = \int dz\, p(z)z \tag{4.17}$$

and

$$M_{II} = \bar{k}_c = \int dz z^2 p(z) \tag{4.18}$$

where c_o has the usual meaning, k_o is the bending rigidity and $\bar{k}_c$ is the elastic modulus for Gaussian curvature. Importantly, M_I and M_{II} are related with experimentally accessible quantities: $c_o k_o$ and $\bar{k}_c$. Otherwise, both moments have been calculated for a variety of phospholipids (Cantor 1999). Depending on the lipid environment, it has been calculated that the pressure profile ranges from 0.8 to 3.3 $k_B T$ (Samuli Ollila et al. 2007). Hence the pressure profile may induce changes in Δw of approx 1–10 $k_B T$. In any event, these studies pointed to a much more wide regulatory effect of lipids on IMPs function (Phillips et al. 2009).

4.7 Identification of Lipids at the Membrane Lipid-Protein Interface

In addition to using the techniques described above to measure lipid compressibility or elasticity and their influence on phospholipids adapting to the surface of an IMP, other experimental approaches can identify the presence of a particular phospholipid species at this interface. An example is FRET between pyrene-labeled phospholipids and IMPs with a single tryptophan engineered into a targeted region. This approach is based on the fact that the pyrene excitation spectrum overlaps extensively with the tryptophan emission spectrum (Fig. 4.15).

In the regular formalism applied to these experiments (Picas et al. 2010; Suárez-Germà et al. 2014) the rate constant for FRET between a donor molecule (D), with fluorescence lifetime τ_0, and an acceptor molecule (A), separated by a distance R, is given by

$$k_T = \frac{1}{\tau_0}\left(\frac{R_0}{R}\right)^6 \tag{4.19}$$

where R_0 is the critical distance (estimated to be 3.0 nm for the Trp/pyrene couple). However, it is important to realize that this equation is only applied to a

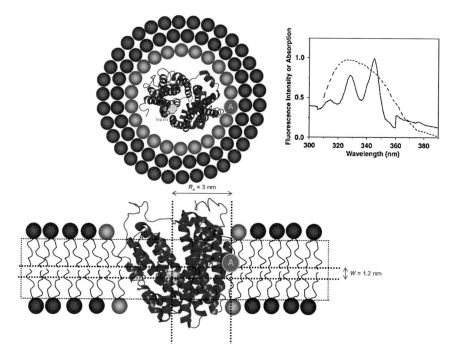

Fig. 4.15 Model of lactose permease embedded in the bilayer: sagital and frontal view showing the distances R, R_e and w between the W151 residue (donor) and the pyrene labelled phospho lipids (acceptor, A). Overlap beween the emission spectrum of LacY mutant single W151/G154C (*dashed line*) and the excitation spectrum of proteoliposomes containing POPG labeled with pyrene in the acyl chain. Adapted from Picas et al. (2010)

single A and D pair, which is not the case in the membrane, where a single donor tryptophan (D) can be surrounded by several A molecules (pyrene labeled phospholipids) (Fig. 4.15). Assuming the A populations deduced from in Fig. 4.15, the fluorescence decay of D molecules can be expressed as

$$i_{DA}(t) = i_D(t)\rho_a(t)\rho_r(t) \tag{4.20}$$

where i_D and i_{DA} describe the fluorescence decays of D in absence and presence of A, respectively with ρ_a and ρ_r being the contributions to the decay of the boundary and randomly distributed bulk acceptors, respectively. Since the number of boundary pyrene phospholipids around each protein molecule is expected to follow a binomial population, their contribution to the decay can be written as

$$\rho_a(t) = \sum_{n=0}^{m} e^{-nk_t t} \binom{m}{n} \mu^n (1 - \mu)^{m-n} \tag{4.21}$$

where m is the number of phospholipid molecules in the first layer surrounding the protein and μ (0 to m sites occupied simultaneously by labeled lipid) represents the probability of each site at the protein-lipid interface being occupied by a labeled pyrene phospholipid. According this model R, the estimation of the distance between the D and A molecules in the boundary region, becomes

$$R = \left(w^2 + R_e^2\right)^{1/2} \tag{4.22}$$

where w is the transverse distance between D and the bilayer centre, and R_e is the exclusion distance along the bilayer plane between the protein axis and the annular lipid molecules. For the LacY-Pyrene phospholipid system, the value $R = 3.2$ nm was used.

The FRET contribution of acceptors randomly distributed outside the annular region is given by

$$\rho_r(t) = \exp\left\{-4n_2\pi l^2 \int_0^{\frac{1}{\sqrt{l^2+R_e^2}}} \frac{1 - \exp\left(-tb^3\alpha^6\right)}{\alpha^3} d\alpha\right\} \tag{4.23}$$

where $b = (R_0/l)^2\tau^{-1/3}$, n_2 is the acceptor density in each leaflet, and l the distance between the plane of the donors and the plane of acceptors.

The probability μ of each of the m annular sites in each leaflet of being occupied by an acceptor depends on the acceptor molar fraction and on a relative selectivity constant (K_S) which quantifies the relative affinity of the labeled to unlabeled phospholipids. Thus μ can be written as,

$$\mu = K_s \frac{n_{pyr}}{n_{pyr} + n_{PL}} = K_s \chi_{pyr} \tag{4.24}$$

where n's are the mole numbers of the labeled (n_{pyr}) and non-labeled (n_{PL}) phospholipids, χ_{pyr} is the mole fraction of pyrene, and K_s is the relative association constant between the labeled and unlabeled phospholipids. It follows that, $K_s = 1$ denotes equal probability of finding acceptors at the lipid-protein region and in the bulk, whereas $K_s = 0$ denotes the absence of acceptor in the lipid-protein interface. This formalism has been used for the calculation of the probability of finding a specific labeled phospholipid (POPE, POPG) around the LacY. The most interesting outcomes were that: (i) both labeled phospholipids can be found at the lipid-protein interface; (ii) the protein recruits phospholipids in the proportion of each species present in the lipid bilayer; and (iii) that POPE is the lipid that seems to specifically fulfill the physicochemical requirements for the best matching between LacY and the bilayer. This last finding is in agreement with cumulative results coming from *in vivo* experiments and it is consistent with a model that suggests the specific insertion of LacY into fluid phases through a specific interaction with POPE, most probably supported by the c_o value of the phospholipid.

4.8   An Integrative Model for Lactose Permease

In conclusion we have deduced that: (i) IMPs insert preferentially into fluid lipid domains; (ii) there is preference between individual IMPs and specific species of phospholipids; and (iii) the intrinsic curvature of the phospholipids represents a physicochemical requirement involved in adequate lipid-protein matching. To rationalize these deductions, we can conceive that LacY that is inserted in the fluid phase provokes modifications in the lateral pressure profile of the phospholipids, leading to their adaptation to the protein surface. This would lead to the step height difference observed between the domains in Fig. 4.11a, resulting from the contribution of the close-packed assemblies above a "distorted" fluid phase and the gel phase. Actually, the step height analysis of the AFM images of PLSs becomes a valuable tool. As shown in the cartoon in Fig. 4.16, three step height differences between domains were established: 0.4, 0.9 and 1.3 nm. This can be rationalised by taking into account that LacY inserts preferentially into the fluid domain from which it protrudes 1.4 nm (see the white spots in the AFM image in Fig. 4.16). Hence, 0.4 and 0.9 nm would correspond to the differences between protruding entities of LacY and the gel and fluid phase, respectively.

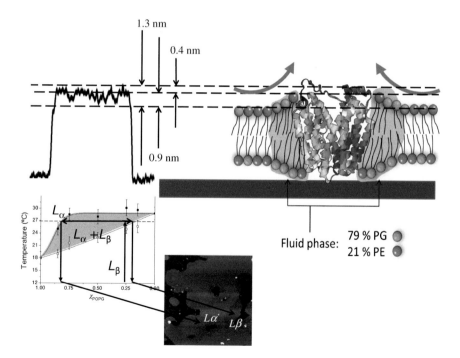

Fig. 4.16 Correspondence between the step height analysis of performed in Fig. 4.11a and hypothetical model of LacY embedded in the bilayers (*top*). Estimation of the composition of the fluid and gel phases on the phase diagram for POPE:POPG by using the lever rule (Fig. 3.10) and the corresponding AFM image (*bottom*)

Since the fluid phase is, according to the phase diagram, constituted by ~79 % of POPG, and knowing from FRET experiments that LacY has preference for POPE, the way to reconcile both findings is to assume that POPE resides mainly around the protein and it is diluted in the bulk formed mainly by POPG.

4.9 A Word on Multidrug Resistance Mediated by Membranes

Of major importance in health is the emergence of resistances to antibiotic and anticancer drugs. Resistance may originate from several mechanisms: drug inactivation, alteration of the specific target, inactivation of the drug or, active efflux of drug mediated by IMPs commonly called multidrug transporters. There are two classes of multidrug transporters: the ABC-type dependent on ATP hydrolysis for pumping out the drugs from the cell, and secondary transporters that extrude drugs via mechanisms that are coupled energetically to the electrochemical potential of H^+ or Na^+. Some secondary multidrug transporters, such as NorA form *Staphyloccocus aureus*, have a secondary structure very similar to LacY. Efflux pump mechanisms are not very well understood but experimental evidence suggests that they depend on the affinity of the drug for the bilayer. Among the mechanistic models proposed for drug resistance the drug always reaches the protein from within the bilayer, suggesting an involvement of the phospholipids in the neighborhood of the multidrug transporters in a coupled mechanism that leads to the drug expulsion from the cytoplasm (Putman et al. 2000).

References

Attard GS, Templer RH, Smith WS, Hunt AN, Jackowski S. Modulation of CTP:phosphocholine cytidylyltransferase by membrane curvature elastic stress. Proc Natl Acad Sci U. S. A. 2000;97(16):9032–6.

Bogdanov M, Heacock P, Guan Z, Dowhan W. Plasticity of lipid-protein interactions in the function and topogenesis of the membrane protein lactose permease from *Escherichia coli*. Proc Natl Acad Sci U. S. A. 2010;107(34):15057–62.

Cantor RS. Lateral pressures in cell membranes: a mechanism for modulation of protein function. J Phys Chem. 1997;101(96):1723–5.

Cantor RS. The influence of membrane lateral pressures on simple geometric models of protein conformational equilibria. Chem Phys Lipids. 1999;101(1):45–56.

Caplan SR, Essig A. Bioenergetics and linear nonequilibrium thermodynamics the steady state. 2nd ed. NY: Harvard University Press; 1983.

Dowhan W, Mileykovskaya E, Bogdanov M. Diversity and versatility of lipid-protein interactions revealed by molecular genetic approaches. Biochim Biophys Acta. 2004;1666(1–2):19–39.

Elmore DE, Dougherty DA. Investigating lipid composition effects on the mechanosensitive channel of large conductance (MscL) using molecular dynamics simulations. Biophys J. 2003;85(3):1512–24.

Goormaghtigh E, Raussens V, Ruysschaert J-M. Attenuated total reflection infrared spectroscopy of proteins and lipids in biological membranes. Biochim Biophys Acta Rev Biomembr. 1999;1422:105–85.

Gullingsrud J, Schulten K. Lipid bilayer pressure profiles and mechanosensitive channel gating. Biophys J. 2004;86(6):3496–509.

Haines TH, Dencher NA. Cardiolipin: a proton trap for oxidative phosphorylation. FEBS Lett. 2002;528:35–9.

Houslay MD, Stanley KK. Dynamics of biological membranes. Chischester: Wiley; 1982.

Jiang Y, Lee A, Chen J, Ruta V, Cadene M, Chait BT, et al. X-ray structure of a voltage-dependent K^+ channel. Nature. 2003;423(6935):33–41.

le Coutre J, Narasimhan LR, Patel CK, Kaback HR. The lipid bilayer determines helical tilt angle and function in lactose permease of *Escherichia coli*. Proc Natl Acad Sci U. S. A. 1997;94(19):10167–71.

le Coutre J, Kaback HR, Patel CK, Heginbotham L, Miller C. Fourier transform infrared spectroscopy reveals a rigid alpha-helical assembly for the tetrameric *Streptomyces lividans* K^+ channel. Proc Natl Acad Sci U. S. A. 1998;95(11):6114–7.

Madden TD, Quinn PJ. Arrhenius discontinuities of Ca^{2+}-ATPase activity are unrelated to changes in membrane lipid fluidity of sarcoplasmic reticulum. FEBS letters. 1979;197(1):110–2.

Marius P, Alvis SJ, East JM, Lee AG. The interfacial lipid binding site on the potassium channel KcsA is specific for anionic phospholipids. Biophys J. 2005;89(6):4081–9.

Merino-Montero S. Dissertation Thesis, University of Barcelona. 2005.

Mitchell P. Coupling of phosphorylation to electron and hydrogen transfer by a chemi-osmotic type of mechanism. Nature. 1961;191:144–8.

Muller DJ. AFM: a nanotool in membrane biology. Biochemistry. 2008;47(31):7986–98.

Ollila S. Lateral Pressure in Lipid Membranes and Its Role in Function of Membrane Proteins. Dissertation thesis. Tampere University of Technology. 2010.

Perozo E, Rees DC. Structure and mechanism in prokaryotic mechanosensitive channels. Curr. Opp. Struct. Biol. 2003;13:432–42.

Phillips R, Ursell T, Wiggins P, Sens P. Emerging roles for lipids in shaping membrane-protein function. Nature. 2009;459(7245):379–85.

Picas L, Montero MT, Morros A, Vázquez-Ibar JL, Hernández-Borrell J. Evidence of phosphatidylethanolamine and phosphatidylglycerol presence at the annular region of lactose permease of *Escherichia coli*. Biochim Biophys Acta Biomembr. 2010;1798(2):291–6.

Prats M, Tocanne JF, Teissié J. Lateral proton conduction along a lipid-water interface layer: a molecular mechanism for the role of hydration water molecules. Biochimie. 1989;71(1):33–6.

Putman M, van Veen HW, Konings WN. Molecular properties of bacterial multidrug transporters. Microbiol Mol Biol Rev. 2000;64(4):672–93.

Samuli Ollila OH, Róg T, Karttunen M, Vattulainen I. Role of sterol type on lateral pressure profiles of lipid membranes affecting membrane protein functionality: comparison between cholesterol, desmosterol, 7-dehydrocholesterol and ketosterol. J Struct Biol. 2007;159 2 SPEC. ISS. :311–23.

Schmidt D, Jiang Q-X, MacKinnon R. Phospholipids and the origin of cationic gating charges in voltage sensors. Nature. 2006;444(7120):775–9.

Seeger HM, Bortolotti CA, Alessandrini A, Facci P. Phase-transition-induced protein redistribution in lipid bilayers. J Phys Chem B. 2009;113(52):16654–9.

Serdiuk T, Madej MG, Sugihara J, Kawamura S, Mari SA, Kaback HR, et al. Substrate-induced changes in the structural properties of LacY. Proc Natl Acad Sci. 2014;111:E1571–80.

Serdiuk T, Sugihara J, Mari SA, Kaback HR, Müller DJ. Observing a lipid-dependent alteration in single lactose permeases. Structure. 2015;23(4):754–61.

Suárez-Germà C, Domènech Ò, Montero MT, Hernández-Borrell J. Effect of lactose permease presence on the structure and nanomechanics of two-component supported lipid bilayers. Biochim Biophys Acta Biomembr. 2014;1838(3):842–52.

Templer RH, Castle SJ, Curran a R, Rumbles G, Klug DR. Sensing isothermal changes in the lateral pressure in model membranes using di-pyrenyl phosphatidylcholine. Faraday Discuss. 1998;111:41–53; discussion 69–78.

Valiyaveetil FI, Zhou Y, MacKinnon R. Lipids in the structure, folding, and function of the KcsA K^+ channel. Biochemistry. 2002;41(35):10771–7.

Vitrac H, Bogdanov M, Dowhan W. Proper fatty acid composition rather than an ionizable lipid amine is required for full transport function of lactose permease from *Escherichia coli*. J Biol Chem. 2013;288:5873–85.

Weingarth M, Prokofyev A, van der Cruijsen EAW, Nand D, Bonvin AMJJ, Pongs O, et al. Structural determinants aspects of specific lipid binding to potassium channels. J Am Chem Soc. 2013;135:3983–8.

Zhang W, Kaback HR. Effect of the lipid phase transition on the lactose permease from Escherichia coli. Biochemistry. 2000;39(49):14538–42.

Printed in the United States
By Bookmasters